HETEROMETALLIC COORDINATION COPPER (II) COMPOUNDS:

CLASSIFICATION AND ANALYSIS OF CRYSTALLOGRAPHIC AND STRUCTURAL DATA

HETEROMETALLIC COORDINATION COPPER (II) COMPOUNDS: CLASSIFICATION AND ANALYSIS OF CRYSTALLOGRAPHIC AND STRUCTURAL DATA

MILAN MELNÍK, GREGOR ONDREJOVIC,
MÁRIA KABESOVÁ, ZUZANA BALOGHOVÁ
AND ANNA BROSKOVICOVÁ

Nova Science Publishers, Inc.
Huntington, New York

Editorial Production:	Susan Boriotti
Office Manager:	Annette Hellinger
Graphics:	Frank Grucci and Jennifer Lucas
Information Editor:	Tatiana Shohov
Book Production:	Donna Dennis, Patrick Davin, Christine Mathosian, Tammy Sauter and Lynette Van Helden
Circulation:	Maryanne Schmidt
Marketing/Sales:	Cathy DeGregory

Library of Congress Cataloging-in-Publication Data

Heterometallic Coordination Copper (II) Compounds: Classification and Analysis of Crystallographic and Structural Data / Milan Milník..[et al.].

p. cm.

Includes bibliographical references and index.

ISBN 1-56072-747-0

1. Organocopper compounds. 2. Coordination compounds. I. Milník, M. (Milan), 1938- .

QD412.C9.H48 1999 99-049906

547'.05652--dc21 CIP

227 Main Street, Suite 100
Huntington, New York 11743
Tele. 631-462-6682 Fax 631-462-4666
e-mail: Novascience@earthlink.net
e-mail: Novascil@aol.com
Web Site: http://www.nexusworld.com/nova

Printed in the United States of America

CONTENTS

LEGEND TO THE FIGURES

HETEROMETALLIC COORDINATION COPPER (II) COMPOUNDS: CLASSIFICATION AND ANALYSIS OF CRYSTALLOGRAPHIC AND STRUCTURAL DATA

Milan Melník*, Gregor Ondrejovic, Mária Kabesová, Zuzana Baloghová, and Anna Broskovicová

Department of Inorganic Chemistry
Slovak Technical University
SL-81237 Bratislava
Slovak Republic

0. ABBREVIATIONS

aapen	3,8-dimethyl-1,10-dio(o-hydroxyphenyl)-4,7-diazadeca-2,8-diene-1,10-dionate
ac	acetate
acac	acetylacetonate
acacP	(o-(diphenylphosphino)-benzoyl)pinacolinate
acpa	N-(1-acetyl-2-propylridene)(2-pyridylmethyl)aminate
acpy	N-acetylacetonylidene
ae	7-amino-4-methyl-5-aza-3-hepten-2-onate
aet	2-aminoethanol
ahmp	2-amino-2-hydroxy-methyl-1,3-propanediolate
[9]aneN_3	1-formyl-4,7-bis(2-hydroxy-2-methyl-propyl)-1,4,7-triazacyclononane
ala	alanine
alaha	L-α-alanine hydroxamate
attn	1,9-bis(3-amino-4H-1,2,4-triazol-5-yl)-3,7-dithianonane
$(baa)_2en$	5,5'-(1,2-ethanediyldinitrilo)bis(1-phenyl-1,3-hexanedionate)
bado	N,N'-bis(3-amino-2,2-dimethylpropyl)oxamidate
bcb	condensation product of 5-tert-butyl-2-hydroxy-1,3-benzenedicarbaldehyde with 1,2-phenylenediamine
bdb	1,4-bis(p-methoxyphenyl)-1,4-diaza-4-methylbutadien-2-yl
bdd	1,3-benzene bis(1,4-dioxanona-6,8-dionate)
bdmap	1,3-bis(dimethylamino)-2-propanol
bdmmp	2,6-bis(dimethyl-aminomethyl)-4-methylphenolate
bdmmpH	2,6-bis(dimethyl-aminomethyl)-4-methylphenol
bet	betaine
$bhppH_2$	1,3-bis(2-hydroxyphenyl-1,3-propanedionate)
bipym	2,2'-bipyrimidine
bispictn	N,N--bis(2-pyridylmethyl)-1,3-propanediamine
bmm	1,2'-bis((3-methoxysalicylidene) diamino)-2-methylpropane

* Author for correspondance

bnpy$_2$	N,N-bis(2-(2-pyridylethyl))benzylamine
bpac	1-benzyl-4,7-diisopropyl-1,4,7-triazacyclononane
bpmp	2,6-bis(-2-pyridylmethylamino)methyl)-4-methylphenol
bpy	2,2'-bipyridine
bzasp	N-(benzyloxycarbonyl)-L-aspartate
cd	ligand derived from condensation of 2,6-diformyl-4-C_2H_4 methylphenol, 1,2-diaminoethane and 1,4-diaminobutane
cdta	cyclohexane-1,2-diamine-N,N,N',N'-tetraacetate
β-cdH	β-cyclodextrin
$C_2H_4O_2$	ethylene glycolate
$C_2H_6O_2$	ethylene glycol
C_5H_4NO	2-pyridonate
C_8H_{10}	ethylbenzene
$C_7H_{12}O_5$	methyl-α-L-rhamnopyranosid-(2,3)-ate(2-)
$C_{10}H_{24}N_4$	2,4-dimethyl-5,8-diazadec-4-ene-2,10-diamine
chp	6-chloro-2-pyridone
cit	citrate(3-)
citH	citrate(2-)
Clac	chloroacetate
Cl_2opba	4,5-dichloro-o-phenylene
Cl_4obbz	oxamide-N,N'-bis(3,5-dichlorobenzoate)
Br_4obbz	oxamide-N,N'-bis(3,5-dibromobenzoate)
cnda	2-carboxyethylnitrilodiacetate(3-)
cp	cyclopentadienyl
cp*	pentamethylcyclopentadienyl
cphi	4-cyanophenylisocyane
cpme	$(\eta^5\text{-}C_5H_4)CH_2NH(CH_2)_2(\eta^5\text{-}C_5H_4)$
cryptand	1,4,12,15,18,26,31,39-octaazapentacyclo$[13,13,13]^{6.10}I^{20.24}$ $I^{33.37}$I-tetracontane-6,8,10,20,22,24,33,35,37- nonaene 85?
csh	2-(5-chloro-salicylideneamino)-1,3-dihydroxy
cth	D,L-5,5,7,12,12,14-hexamethyl-1,4,8,11-tetraaza-cyclotetradecane
cyclam	1,4,8,11-tetraazacyclotetradecane
dat	6,11-dimethyl-7,10-diazahexadeca-5,11-diene-2,4,13,15-tetraonate
dbs	condensation product of 2,3-diamino-2,3-dimethylbutane with 5-tertbutyl-3-pyridyl-salicylaldehyde
deae	2-(diethylamino)-2-propanolate
det	2-diethylenetriamine

dced	1,2-dicyanoethylenedithiolate
ddc	3,4,9,10-dibenzo-1,12-diaza-5,8-dioxacyclopentadecane
ddn	1,12-dodecanedinitrile
dgl	diglycolate
dien	diethylenetriamine
diena	N-(2-aminoethyl)-1,2-ethanediamine
diene	bis(2-aminoethyl)amine
dm	two 2,6-bis(iminomethyl)-4-methyl-phenolates combined with two chains CH_2CH_2 and $CH_2CH_2SCH_2CH_2$
dmap	1,3-bis(dimethylamino)-2-propanolate
dmae	2-dimethylaminoethanol
dmf	dimethylformamide
dmg	dimethylglyoximate
dmit	isotrithionedithiolate
dmim	1,2-dimethylimidazole
dmso	dimethylsulfoxide
dopn	3,9-dimethyl-4,8-diazaundeca-3,8-diene-2,10-dione dioximate
dpb	1,8-bis(2,8,13,17-tetraethyl-3,7,12,18-tetramethylporphyrin-5-yl)biphenylene
dpe	1,1-di(2-pyridyl)ethyl-methyl-ether
(dpe)py_2	α,α-5,15-bis(o-(nicotinoylamino)phenyl)-2,8,12,18-tetraethyl-3,7,13,17-tetramethylporphyrinate
dppn	2,6-bis(2'-pyridyl)pyridazine
dps	2-((diphenylphosphino)-amino)ethyl)salicylaldiminate
dtma	4- diethylenetriamineacetate
dtp	2,2'-thiodiphenolate
edds	N,N'-ethylenediamine-disuccinate
edta	ethylenediamine tetraacetate
en	ethylenediamine
eob	3,3'-[1,2-ethanediyl bis(oxy-2,1-ethanediyloxy)]bis(2-hydroxybenzaldehyde -O-methyloximate)
epn	2,2'-(1,2-ethanediyl bis(nitrilomethylidene)) bis(pyridine N-oxide)
eryan	methyl-α-D-mannopyranosid
eth	ethyleneglycolate(2-)
$EtSiO_2$	ethylsiloxanolate
F_3ac	trifluoroacetate
$(fsa)_2en$	N,N'-(2-hydroxy-3-carboxybenzylidene)-1,2-diaminoethanate
fsaaep	derivate from 3-(N-2-(pyridylethyl)formimidoyl)salicylate

F_8tpp	tetrakis(2,6-difluoro-phenyl)porphyrinate(2-)
fms	binucleating macrocycle comprised of 2,6-diformyl-5-methyl phenol, ethylenediamine and 1,3-diaminopropan-2-ole
gly	glycine
ghg	glycyl-L-histidylglycinate
Hapen	N,N--ethylene bis(o-hydroxyacetophenone iminate)
hbcpa	N(o-hydroxy-benzylidene-cyclopentadienyleneamine)
hdbc	9,10,12,13,15,16-hexahydro-3,7,18,22-dimetheno-8,11,14,17,1,24-benzotetraoxadiazacyclohexacosine-29,30-diolate
hfac	hexafluoroacetonate
hfacac	hexafluoroacetylacetonate
hmpa	hexamethylphosphoramide
hoat	1,4,7,16,19,22-hexaza-10,13,25,28-tetraoxacyclotriacontane
hpctp	hexakis(3-pyrazolyl)cyclotriphosphane
ic	isocyanurate
im	imidazole
mc	macrocycle formed by the condensation of two molecules of 2,6-diformyl-4-methylphenol and two diamines
mch	macrocycle with two 2,6-bis(iminoethyl)-4-methylphenolate entities combined through two luteral chains -$(CH_2)_2$- and -$(CH_2)_2(NH)(CH_2)_2$-
mec	1-methylcytosine
Me_2oxpn	2,2'-dimethyl-N,N'-bis(3-aminopropyl)oxamide
4-Mepy	4-methylpyridine
Me_2pz	3,5-dimethylpyrazolyl
$Me_3[10]N_3$	2,4,4-trimethyl-1,5,9-triazacyclodec-1-ene
Me_5dien	1,1,4,7,7-pentamethyldiethyllenetriamine
Me_6tren	2,2'2"-,tris(2-(dimethylamino)ethyl)amine
$Me_3[12]N_3$	2,4,4-trimethyl-1,5,9-triazacyclododec-1-ene
Me_6-[14]ane-N_4	(±)-5,7,7,12,14,14-hexamethyl-1,4,8,11-tetraazacyclotetradecane
mesalen	N,N'-ethylenebis(2-hydroxyacetophenoimine)
meu	1-methyluracilate
mhp	6-methyl-2-pyridone
mim	N-methylimidazole
mnt	maleonitriledithiolate
modeH	methoxydietoxidate
mode	methoxydiethoxide
moe	2-methoxyethoxidate
mopH	3-methoxypropanol
mop	3-methoxypropanolate
mp	3-methyl-2-pyrrolidone
mph	6-methyl-2-pyridone

moim	4-(6-methyl-8-oxo-2,5-diazanonane-1,5,7-trienyl)imidazole
mol	2-methoxyethanolate
mpz	3-methylpyrazolyl
mtfd	2-methyl-2,4,6-tris(trifluoromethyl)1,3-dioxane-4,6-diolate
mtm	(bis(methylthio)-methylidene)malonate
niox	cyclohexane-1,2-dione-dioximate
n-15-crown-5	naphto-15-crown-5
nup	bis((1,3-dimethyl-5-nitrouracil-6-yl)-amino)propane
nta	nitrilotriacetate
ntac	nitriloacetate
obbz	oxamidobis(benzoate)
obze	oxamido-N-benzoato-N'-ethanoate
obzp	oxamido-N-benzoato-N-propionate
od	meso-oxalandiolate
odp	1K5C-1-oxo-1,2-diphenyl-3,3-dicarbbutoxy-5-carboxy-2KO7-1,2-diphospha-1KP2-cyclopentan-4-one
oep	octaethylpophyrine
opb	oxamidobis(propionate)
opba	o-phenylenebis(oxamate)
or	orotate
ox	oxalate
oxap	N,N'-bis(2-aminopropyl)oxamide(2-)
oxen	N,N'-bis(2-aminoethyl)oxamide
oxpn	N,N'-bis(3-aminopropyl)oxamide
oxpno	oxamidate
paba	arylazo oxamidate
pba	1,3-propylene-bis(oxamate)
pbaOH	2-hydroxo-1,3-propylenebis(oxamate)
pdac	o-phenylenediamine-N,N,N',N'-tetraacetate
phb	phenobarbiturate
phen	1,10-phenanthroline
pht	o-phthalate
pmdt	1,1,4,7,7-pentamethyldiethylene-triamine
pnaoH	2,2'-(1,3-diamino-propane)bis(2-methyl-3-butanone)oximate
pr	propionate
$(prp)_2en$	Schiff base ligand prepared from the condensation of o-hydroxypropiophenone with ethylenediamine
$(prp)_2pr$	Schiff base ligand prepared from the condensation of o-hydroxypropiophenone with 1,3-propylenediamine
pr^iO	izopropylalkohol
prp	propiophenone
pvo	pyruvamide oximate
py	pyridine
2-pyc	2-pyridinecarboxylate
pyH	pyridinium

pyo	2-hydroxy-pyridine(-)
pyor	pyrrolidine-2-one
pz	1-pyrazolyl
pza	pyrazinate
saladhp	1,3-dihydroxy-2-methyl-2-(salicylideneamine)propanate(3-)
salal	N-salicylidene-β-alaninate
salen	N,N-ethylene bis(salicylaldiminate)
sb	Schiff base from 1-(N,N-dimethylaminomethyl)-2-formylcymantene and (s)-alanyl-(s)-alanine
salphen	ligand include derivatives of the macrocyclic tetraazacyclo-tetradecacatechol and of the salicylideneaminocatechols obtained by condensation of 4,5-diaminocatechol with salicylaldehydes
salpn	N-(3-aminopropyl)salicylaldiminate
saltn	N,N-1,3-propylene bis(salicylaldiminate)
$S_2C_2O_2$	dithiooxalate
sgl	squarate
tacac	1,4,7-triazacyclo-nonane-N-acetate
tact	1,4,8,11-tetraazacyclotetradecane
tba	tetrabenzo[b,f,j,n][1,5,9,13]tetraaza cyclohexadecine
tdeH	2,2'-thiodiethanolate
tem	38-membered macrocycle having eight amine and four phenol functionalities
terpy	2,2':6',2"-terpyridine
thd	2,2,6,6-tetramethyl-3,5-heptadionate
thf	tetrahydrofuran
tmac	1,4,7-trimethyl-1,4,7-triazacyclononane
tmeda	N,N,N',N'-tetramethyl-ethylenediamine
tmpa	tris(2-pyridylmethyl)amine
tpp	5,10,15,20-tetraphenylporphyrinate
tren	tris(2-aminoethyl)ammine
trien	triethylenetetramine
trf	triflate
tris	2-amino-2-hydroxymethyl-1,3-propanediolate
trisH	2-amino-2-hydroxymethyl-1,3-propanediol
tsgly	N-tosyl-glycinate
ttfch	1,3,5-tris(trifluoromethyl)-1,3,5-trioxycyclohexane
wil	willardiine [1-(2-amino-2-carboxyethyl)uracil]

1. INTRODUCTION

The chemistry of copper compounds is an extensive and active area of study, with the relationships between structure, reactivity and catalytic activity being of major importance. There have been many structural studies of copper compounds. Copper compounds have been summarised and classified from a structural point of view: copper(I) [1], copper(II) [2-6], copper(III) [7] and mixed-valence copper(I)-copper(II) derivatives [8]. A comprehensive review of heterometallic copper(I) compounds has been also published [9], but there has not been a such review of heterometallic copper(II) compounds to date. The aim of this review is to filt this gap by drawing together all of the heterometallic copper(II) complex structures and structural data, from X-ray and netron diffraction, up to the end of 1997.

There are almost four hundred derivatives and these have been classified according to their nuclearity. Copper(II), although classified as a "borderline" acceptor atom, is very adaptable and is found with a variety of "soft", "borderline" and "hard" ligand donor atom.

2. HETEROMETALLIC DOUBLE SALTS OF COPPER(II)

Crystallographic and structural data for the double salts are given in Table 1. The compounds have been listed and referenced in the order of increasing coordination number about the copper(II) atom. Deep blue compound contains discrete $[Cu(C_2H_4O_2)_2]^{2-}$ and $[Ba(C_2H_4O_2)_6]^{2+}$ units [10]. Copper(II) is almost square-planar in its oxygen coordination with Cu-O bond lengths range from 1.912(4) to 1.934(4)Å and O - Cu - O bond angles from 86.5(2) to 93.6(2)°. Barium(II) atom is coordinated by nine oxygen atoms (the average Ba-O bond lenght is 2.82 Å); six oxygen atoms are from bidentate ethylene glycol ligands, and the remaining three oxygen atoms come from three monodentate ethylene glycol ligands. The bonding environment around barium is a tricapped trigonal prism. The Cu - containing and Ba - containing moieties are interconnected by hydrogen bonding to form chains of alternating $Cu(C_2H_4O_2)_2^{2-}$ and $Ba(C_2H_6O_2)_6^{2+}$ units.

Structure of deep-blue Millon's salt [11] consists of $Cu(NH_3)_4^{2+}$ and $PtCl_4^{2-}$ square-planar units alternating in stacks along the c axis with 2.00 (1) Å Cu-N and 2.298 (4) Å Pt-Cl separations.

Red salts of the composition $[Cu(en)_2][Pt(en)_2X_2](ClO_4)_4$ (X = Cl or Br] [12] contain a square-planar coordination about the copper(II) by two chelated ethylenediamine molecules and an pseudo-octahedrally coordinated Pt(IV) atoms with the chromophore PtN_4X_2. Bromine salt exists in two isomeric forms, differ mostly by degree of distortion (Table 1).

Orange derivative [13a] contains $[Cu(papH)]^-$, $[Li(H_2O)_4]^+$ and two solvent water molecules. The papH in the complex anion is coordinated in a tetradentate fashion, forming three condensed chelate rings, two five- membered with an average N - Cu - N angle of 82.2(2)° and one six-membered with N - Cu - N angle of 98.6(2)°. The mean Cu - N (amide) bonds of 1.918(4)Å is about 0.040Å shorter than those of Cu-N(oxime) ones (1.958(4)Å).

X-ray analysis of $[Cu(oep)]_2SbCl_6.0.5CH_2Cl_2$ shows [13b] that the two rings of the $[Cu(oep)]_2$ dimers are crystallographically equivalent with Cu - Cu distance of 3.466(2)Å . Each Cu(II) atom is in a square-planar environment with an average Cu - N bond distance of 1.984(6)Å . Deviations of Cu(II) atom from the planarity in the unique porhyrine core is 0.07Å . The Sb(V) atom is octahedrally coordinated by six chlorine atoms.

Grass green salt [14] composed of well separated $[CuCl_4]^{2-}$, $[Pt(en)_2 Cl_2]^{2+}$ and water molecule. The $[CuCl_4]^{2-}$ anion is situated on the two-fold axis through (x,x) = (1/2, 1/4). There are two independent Cu - Cl distances, 2.239(5) (x2) and 2.276(4)Å (x2). The Cl - Cu - Cl angles (Table 1) show a large distortion of the $[CuCl_4]^{2-}$ anion from tetrahedral towards square-planar configuration. The platinum (IV) atom lies at a centre of symmetry which causes the four nitrogen atoms of two ethylenediamine molecules and the Pt(IV) to be exactly in a plane and the chlorines to be in *trans* position.

Brown salt [15] is composed of three fragment, two symmetry independent $[Sc(15\text{-crown-}5)Cl_2]^+$ cations and a $[CuCl_4]^{2-}$ anion, all of which are situated in special positions on mirror planes. The scandium(III) atoms in the cations are almost exactly at the centres of the cavities of the macrocycle 15-crown-5 molecules and are located in the plane formed by the five oxygen atoms in the molecule and two chlorines in apical positions about each Sc(III) atom completed an almost ideal pentagonal bipyramid.

The $[CuCl_4]^{2-}$ anion has the distorted (flattened) tetrahedral shape (Table 1).

There are two brown salts $[CuCl_4]$ $[M(15\text{-crown-}5)(H_2O)_2]$ (M = Mg [15] or Co [17]), in which M(II) atoms are in a pentagonal-bipyramidal environment with two water molecules in apical positions. The $[CuCl_4]^{2-}$ anions have the distored tetrahedral arrangement.

Another two orange salts of the composition $[Cl_2Cu(\mu\text{-}Cl)_2CuCl_2][M(15\text{-crown-}5)(MeCN)_2]$, (M = Mg [16] or Co [17]) are isostructural. The copper(II) atom tetrahedra are strongly distorted in the centrosymmetrical anion $[Cu_2Cl_6]^{2-}$ and bonded through two μ-chlorines. The bond angles vary in a wide range from 86.6 to 144.3°. Mean values of the terminal Cu - Cl bonds (2.19 Å) are noticeably shorter than those of bridging chlorines 2.31 Å). The M(II) atoms are surrounded by five oxygen atoms of 15-crown-5 macrocycle molecules in plane and two MeCN molecules (at apical positions) completed a pentagonal-bipyramidal environment about each M(II) atom with the chromophore MO_5N_2 (Table 1).

Brown-green crystal contains [17b] the complex cations $[K(\text{naphtho-15-crown-}5)_2]^+$, the anion $[Cu_2Cl_6]^{2-}$ and water molecules. The coordination number of potassium(I) is equal to 10, the coordination polyhedron is a pentagonal antiprism. In dimeric anion, $[Cu_2Cl_6]^{2-}$, two of the chlorine atoms act as bridges for two copper(II) atoms and the Cu - Cl distances range from 2.314(6) to 2.352(6)Å. The remaining four chlorine atoms are terminal with Cu - Cl distances range from 2.157(6) to 2.177(6)Å. The copper(II) atoms are located at distance equal to 3.331(3)Å. The coordination polyhedron is a distorted tetrahedron with the Cl - Cu - Cl angles ranging from 88.5(2) to 132.8(2)°.

Black salt [18] contains $[Cu(mnt)_2]^{2-}$ anion and two $[Fe(cp^*)_2]^+$ cations. In complex anion, a square planar environment about the copper(II) atom is created by two chelated maleonitriledithiolates with the chromophore CuS_4 (Table 1). Iron atoms in the complex cations are sandwiched by decamethylcyclopentadienyls.

In another three derivatives [19] two chelate bidentate lysine ligands created a square-planar environment about the copper (II) atoms with the chromophore CuO_2N_2.

In yellow salt [20], the Cu(II) atom is coordinated by pairs O and N atoms of a tetradentate hdbc ligand, forming three condensed chelate rings, one five-membered with N - Cu - N bond angle of 84.7(2)° and two six-membered with an average O - Cu - N bond angle of 93.7(2)°. The mean Cu - O bond distance of 1.896(5)Å is about 0.040Å shorter than that of Cu - N bond distance (1.936(5)Å). The iron atom is coordinated by one η^5-cyclopentadienyl molecule, two CO groups and one NH_3 molecule (Table 1).

X-ray analysis and neutron diffraction of $[CuCl_5][Co(NH_3)_6]$ shows [21], that the copper(II), atom is coordinated by five chlorine atoms in the fashion of a compressed trigonal bipyramid with the mean Cu - Cl (axial) and Cu - Cl (equatorial) bond distances of 2.291(1) and 2.392(2)Å, respectively. The cobalt (III) atom is surrounded by six NH_3 molecules in a regular octahedral environment with Co - N bond distance of 1.968(2)Å.

There are three orange salts of the composition $[Cu_2Cl_8][M(en)_3]Cl_2.2H_2O$; (M(III) = Co [22], Ir or Rh [23]), which are isostructural. The $Cu_2Cl_8^{4-}$ anion consists of two distorted trigonal bipyramids sharing a common edge with the mean Cu - Cl(equatorial) bond distance of 2.285Å and two non-aquivalent Cu - Cl(axial) bond distances of 2.316 and 2.704Å. All M(III) atoms are in a rhombic - octahedral environment with the chromophore of MN_6.

Green salt of the composition $[Cu(trisH)(tris)(H_2O)]_2CrO_4$ contains [24] two crystallographically independent molecules. Each copper(II) atom have a square-pyramidal coordination sphere with two amino and two hydroxymethyl groups in the basal plane and a water molecule as the apical ligand. The copper(II) atoms deviates from the plane towards the apex of the pyramid by 0.1898(4) and 0.1873(4)Å for molecules 1 and 2, respectively. The five- membered chelate rings have an envelope conformation with the mean O - Cu - N angles of 84.8(1)°. The chromium(VI) atoms are tetrahedrally coordinated by oxygen atoms (Table 1).

Brown salt of the composition $[(bpac)Cu(\mu\text{-}O)_2Cu(bpac)](SbF_6)_2.7Me_2CO$ .2MeCN [25] was studied also by X-ray analysis. The $[\{(bpac)Cu\}_2(\mu\text{-}O)_2]^{2+}$ cation consists of two distorted square pyramids shaving a common edge with the mean Cu - O(bridge) distance of 1.806Å and with Cu - Cu distance of 2.794(2)Å. All Sb(V) atoms are surrounded by six fluorine atoms.

Structure of blue salt [26] is built up of $[Cu(H_2O)_6]^{2+}$ octahedrons and $[ZrF_8]^{4-}$ square antiprisms linked by hydrogen bonds.

X-ray analysis at 100 K of $[Cu(\mu\text{-}ddn)_3][SbCl_6]_2$ shows [27] a two-dimensional network, in which the copper(II) atoms are linked in two different ways by the 1,12-dodecanedinitrile ligands. The coordination polyhedron around copper(II) shows Jahn-

Teller deformation and consists of six N atoms (Table 1). The space between layers is filled by $SbCl_6^-$ units.

In black salt [28] the copper(II) atom on the inversion center is bound in distorted octahedral fashion of which four nitrogen atoms from 1,4,8,11-tetraazacyclotetradecane and two nitrogen atoms from MeCN occupy the equatorial and axial sites, respectively. Bond distances between copper(II) and coordinated atoms are 2.028(7) - 2.019(7)Å for equatorial and 2.491(6)Å for axial nitrogen atoms. The axial bond is tilted 2.1(2)° with respect to the right angle of the equatorial plane. The Ni(II) atom in the $[Ni(dmit)_2]^-$ anion is coordinated by four sulphur atoms, and the coordination geometry about the nickel atom is a square plane.

Structure of blue green salt [29a] is made up of $[Cu(\mu\text{-bipym})_2(H_2O)(SO_4)_3]^{2-}$ anionic zig-zag chains, $[VO(H_2O)_5]^{2+}$ cations and crystallisation water molecules. A view of the asymmetric dimeric unit within the chain and of a symmetry-related fragment is depicted in Fig. 1. A regular alternation of two non-equivalent copper(II) atoms [Cu(1) and Cu(2)] occurs in the chain. Both exhibit a distorted *cis*-octahedral CuN_4O_2 environment. This chromophore involves a *trans* elongation. The four nitrogen atoms of a bridging bipym ligand occupy equatorial positions of two octahedra [N(1) and N(3) at Cu(1) and N(2) and N(4) at Cu(2)], the two remaining equatorial sites being filled by oxygen atoms from sulfate group [O(2) at Cu(1) and O(3) at Cu(2)] and nitrogen atoms from a second and a third bipym group [N(7) at Cu(1) and N(5a) at Cu(2)]. Six-coordination is achieved by means of the remaining nitrogen atoms from the second and third bipym molecules [N(6) and N(8a) for Cu(1) and Cu(2), respectively] and two oxygen atoms one from a sulfate anion [O(5) at Cu(1)] and the other from a water molecule [O(13) at Cu(2)] (Table 1). Two different bis (chelating) bipym ligands alternate regularly in the chain leading to two intrachain Cu - Cu separations 5.505(2) and 5.675(2)Å for Cu(1)...Cu(2) and Cu(1)...Cu(2), respectively.

In pale brown $[Cu(dpe)_2][Ag_2I_4]$ [29b] the Cu(II) atom of the $[Cu(dpe)_2]^{2+}$ cation lies on a centre of symmetry and is octahedrally coordinated by 1,1-di(2-pyridyl)-ethylmethyl ether through two N and one O donor atoms (Table 1). The centrosymmetric anion displays two tetrahedrally AgI_4 moieties bridged by two iodine atoms.

Red violet salt [30] contains well separated $[Cu(tact)Cl]^+$ cations and $[CdCl_4]^{2-}$ anions. In complex cations, each Cu(II) atom is bonded to one 1,4,8,11-tetraaza-cyclotetradecane molecule and two chlorine atoms which are acting as bridging ligands and connect the copper(II) atoms to chains of distorted octahedra. Distorted tetrahedra of $[CdCl_4]^{2-}$ are situated in cavities between these chains.

Dark brown salt [16] contains $[Cu_3Cl_8(H_2O)_2]^{2-}$ anion and $[Mg(H_2O)_6]^{2+}$ cation. Central copper(II) in the trimetallic anion is in the centre of symmetry and bound by double chlorine bridges with the outer copper atoms and two water molecules. The outer copper(II) atoms are in additional coordinated by two terminal chlorines which completed a tetrahedral environment (Table 1). Magnesium(II) atom is surrounded by six water molecules.

Red brown salt [16] is composed from $[Cu_3Cl_8]^{2-}$ anion and $[Mg(15\text{-crown-}5)\text{-}(MeCN)_2]^{2+}$ cation. Central copper(II) in the trimetalic anion is in the centre of symmetry and bound by triple chlorine bridges with the outer copper(II) atoms. The outer copper(II) atoms are in a tetragonal-pyramidal environment (Table 1). Magnesium(II) atom is located inside the cavity of 15-crown-5 forming coordinate bonds with five oxygens. The polyhedron of magnesium(II) is a pentagonal bipyramid, the axial positions being occupied by the acetonitrile nitrogens.

The data in Table 1 shows copper(II) with four-, five- and six- coordination. The mean value of the Cu-L distance increases with increasing covalent radius of the ligand atom. Thus for four coordination the values are 2.00Å (N, 0.75Å) < 2.24Å (Cl, 0.99Å) for monodentate ligands; 1.92Å (O, 0.73Å) < 2.04Å (N, 0.75Å) < 2.27Å (S, 1.02Å) for bidentate ligands. For five coordination the values are: 2.30Å (O, 0.73Å) < 2.32Å (Cl). The mean Cu - Cl(terminal) distance is shorter than that of Cu - Cl(bridge) with the values of 2.24 vs 2.30 Å. The mean Cu - Cl(bridge) distance increases with coordination number in the sequence:
2.30 Å (four-) < 2.33 Å (five-) < 2.54 Å (six-coordination).

3. HETEROBINUCLEAR COPPER(II) COMPOUNDS

Crystallographic and structural data for these derivatives are gathered in Table 2 and listed in order of increasing Cu(II) - M distance. X-ray structure of red (4-Mepy)-

Cu(μ-ac)$_2$(μ-CO)$_2$Mo(η^5-cp) reveals [31] that the two metal atoms are linked by direct Cu - Mo bond (2.558(1)Å and are also bridged by two acetate units and two CO groups. This is the shortest Cu(II) - M distance found in the heterobinuclear copper(II) complexes.

In blue derivative [32] the Cu(II) and Pt(II) have square-planar coordination spheres and are linked through two 1-methyluracilate ligands in a head-head arrangement, binding to Cu(II) via O and to Pt(II) via N atom. The coordination spheres are completed by two water molecules (Cu) and two NH_3 ligands (Pt), each in *cis* arrangements. The Cu - Pt distance is 2.765(3)Å.

Blue violet (acac)Cu(μ-OSiMe$_3$)$_2$Al(OSiMe$_3$)$_2$ contains [33a] two crystallographically independent heterobinuclear Cu(μ-OSiMe$_3$)$_2$Al bridge is strongly asymmetric with Cu - Al distances of 2.778(4) and 2.784(4)Å and an average Cu - O - Al bridge angles of 94.3(4) and 94.2(4)°, respectively. .Both metal atoms are four-coordinated, CuO_4 (square-planar) and AlO_4 (tetrahedral) (Table 2).

X-ray analysis of CuNi derivative at 123 K shows [33b], that the two fragments, Cu(salen) and Ni(hpac)$_2$, are held together through the phenolic oxygen atoms. The CuO_2Ni network is rigorously planar with Cu - O - Ni bridging angles of 93.9(1)° and Cu - Ni distance of 2.897(2)Å. The copper(II) atom is in a square-planar environment, CuO_2N_2 and the nickel(II) atom has a sligtly distorted octahedron, NiO_6.

In yellow-green (EtOH)$_2$Cu(μ-dbs)Ni [33c], the Ni(II) atom is located in dbs O_2N_2 cavity and exhibits a square-planar coordination geometry, while the Cu(II) atom occupies the outer O_2N_2 binding site with two ethanol molecules as the axial ligands in an octahedral coordination environment. The two metal atoms are bridged by two phenolic oxygen atoms to show a planar Ni - O - Cu - O four- membered ring while the two aromatic spacers are close to coplanar. The Cu - Ni distance is 2.898(2)Å.

Structure of deep red Cu{(μ-baa)$_2$en}Ni [34] consists of heterobinuclear molecules in which the copper(II) atom is bound to four ketonic oxygens and two nitrogens from the ethylenediamine moiety. The intramolecular Cu - Ni distance is 2.925(1)Å.

In green (H_2O)Cu(μ-aapen)Ni [35] the aapen ligand employed has two different coordination sets of atoms, O_2N_2 and O_2O_2, two oxygen atoms being common to both donor sets. In the complex, the nickel(II) atom is planarly retained in the inner O_2N_2

chamber whilst the copper(II) atom, incorporated in the outer O_2O_2 chamber, is five-coordinate, being axially bonded to a water molecule (Table 2). The Cu - Ni distance of 2.931(2)Å is somewhat longer than that found in the red derivative [34].

Heterobinuclear derivative of the composition (MeOH)Cu{μ-(fsa)$_2$en}VO, exists in two isomeric forms, red and dark blue [36]. The structure is made of heterobinuclear units, in which a copper(II) atom is five fold coordinated to two nitrogens, two phenolic oxygens, and the oxygen of a methanol molecule, and a vanadium(IV) atom is also fivefold coordinated to two phenolic and two carboxylic oxygens and the oxygen of the vanadyl group. The structures differ mostly by degree of distortion (Table 2). The Cu - V distance of 2.949(3)Å found in the red derivative is about 0.040Å shorter than that found in the dark blue derivative.

Structure of red CuZn derivative [37] is shown in Fig. 2. The molecular structure consists of a four- coordinate copper(II) atom and a six- coordinate zinc(II) atom. The square and octahedron share one edge formed by the two phenolic oxygen atoms of the mesalen moiety. Two CH_2Cl_2 molecules are packed with the binuclear molecules in the crystalline structure (Fig. 2). The mean Cu - O - Zn bond angle is 93.8(2)° with Cu - Zn distance of 2.966 Å.

In [Cu{μ-(fsa)$_2$en}Ni(H_2O)$_2$].H_2O [38] the copper(II) atom which occupy the inside site exhibits a strict square-planar environment (CuO_2N_2), and the nickel(II) atom occupying the outside site exhibits a pseudooctahedral coordination (NiO_6). The mean Cu - O - Ni bond angle is 98.9(1)° and Cu - Ni distance is 2.9749(6) Å.

X-ray analysis of dark green Cu(μ-dat)VO derivative shows [39] that while copper(II) atom is in a square-planar environment (CuO_2N_2), the vanadium(IV) atom is five- coordinated (VO_5), with Cu - V distance of 2.985Å and the mean Cu - O - V bond angle of 100.3(2)° (Table 2).

In dark maroon derivative [40] cavities in an octadentate-O_2N_6 cypim macrocycle are occupying by two non-equivalent metal atoms, the copper(II) and cobalt(III) atoms. The former has a trigonal-bipyramidal and the latter a pseudo-octahedral environment (Table 2). Two metal moieties, CuN_3O_2 and CoN_4O_2 are held together double by two O donor atoms with the mean Cu - O - Co bond angle of 101.9(4)° and Cu - Co distance of 2.994(2) Å.

In another several derivatives [41-53] a CuO_2M networks are almost planar with Cu - O - M bridging angles ranging from 95.7 to 103.5° and Cu - M distances from 3.001(4) to 3.355(3)Å (Table 2).

In $[Cu(\mu\text{-bpmp})(\mu\text{-pr})_2Fe](PF_6)_2$ [54] the two metals are bridged by the phenolate oxygen atom of the $bpmp^-$ ligand (Cu - O - Fe = 119.3(5)°) and by two propionate groups in a *syn-syn* arrangement. The Cu(II)...Fe(III) separation of 3.401(4)Å ruled out a direct metal-metal bond. Both metal atoms are a pseudo-octahedrally coordinated (MO_3N_3) (Table 2).

X-ray analysis of the CuIr derivative showed [55] that both dppn and nitrosyl ligands bridge the metal atoms with Cu(II) - N - Ir(III) bridge angle of 117(1)° and Cu...Ir separation of 3.416(5)Å.

In $[Cu(\mu\text{-bmm})Gd(NO_3)_3]Me_2CO$ [56], the central region of the structure is occupied by the copper(II) and gadolinium(III) atoms which are bridged one to the other via the two phenolic oxygen atoms the bmm ligand. The two Cu - O - Gd angles differ with values of 105.9(2) and 107.4(2)°. The Cu...Gd separation is 3.428(1)Å.

The asymmetric unit of red CuNi derivative [57a] contains two crystallographically independent heterobinuclear cations, $[(ClO_4)Cu(\mu\text{-nup})\text{-}Ni(Me[12]N_3)]^+$, non-bonded perchlorate anion and two lattice water molecules. Within each cation, copper(II) and nickel(II) are bibridged by two nitroso-oximate groups, one diatomic *syn-syn* and one O- monoatomic, from the pyrimidine and imidazole rings, respectively. This results in intramolecular Cu...Ni distances of 3.440(4) and 3.452(4)Å, for molecules I and II, respectively.

Blue $[(H_2O)Cu(\mu\text{-ala})_4Y(H_2O)_4](ClO_4)_5.3H_2O$ [57b] exists as binuclear molecules with copper(II) and yttrium(III) atoms connected by four bridging carboxylate groups from four alanine molecules in *syn-syn* arrangement. The Cu(II) atom is coordinated by one oxygen atom from a water molecule and four oxygen atoms from the four bridging carboxylate groups, giving a square-pyramidal coordination polyhedron. The Y(III) atom has a square-antiprism environment (YO_8) (Table 2). The Cu...Y separation is 3.491Å.

In brown red CuV derivative [57c] the $aapen^{4-}$ ligand employed has two different coordination sets of atoms, O_2N_2 and O_2O_2 two oxygen atoms being common to both donor sets. In the complex the Cu(II) atom, which is retained in the

O_2N_2 chamber, is five coordinate being axially bonded to a dmso molecule, whilst the uranyl ion is incorporated in the outer O_2O_2 chamber. Another molecule of dmso is retained to preserve the preferred a sligthly distorted pentagonal bipyramid. The Cu...U separation is 3.51Å.

In dark green CuPr derivative [58] the $Pr(hfac)_3$ unit is linked to the copper(II) monomer through a hydroxo and phenoxo bridge with a Cu...Pr distance of 3.538(2)Å. The Cu(II) center has a trigonal bipyramidal geometry (CuO_3N_2) and the Pr(III) center has an irregular polyhedral geometry (PrO_8N).

Structure of deep brown CuMn derivative [59] consists of six- coordinate Mn(III) center, MnO_3N_3, and the copper(II) center has an CuN_4O donor set. The metal moieties are oximato- and acetato- bridged with a Cu...Mn separation of 3.539Å.

In dark red CuFe derivative [60] two square-pyramidal moieties CuN_3O_2 and FeN_4O, are held together by hydroxo bridge with Cu(II) - O - Fe(III) bridge angle of 175.3° and Cu...Fe separation of 3.570Å.

In purple CuRe derivative [61], two non-equivalent metal units a square-pyramidal Cu(II)N_4O and a tetrahedral Re(VII)O_4 are bridged by O atom (Table 2). In dark green CuNi derivative [62], two equivalent a slightly distorted octahedrons, MN_6, are held together by two 1,1-μ-azide groups, with Cu - N - Ni bridge angle of 95.1(1)° and Cu...Ni separation of 3.59(1)Å.

Structure of red purple CuFe derivative [62] is similar to that of dark red CuFe derivative [60]. Two square-pyramidal moieties, CuN_4O and FeN_4O, are bridged by a single O atoms, with Cu(II) - O - Fe(III) bridge angle of 178.2(4)° and Cu...Fe separation of 3.596(2)Å.

In yellow CuMg derivative [15] two non-equivalent units, $CuCl_4$ and MgO_5NCl are held together by a single μ-chlorine atom (Table 2).

In red heterobinuclear CuNi derivative [64] the imidazolate ligand serve as a bridge between two square-planar moieties, MN_3O, with Cu...Ni separation of 3.632(1) Å.

There are two CuBa derivatives, dark brown and green [65], each copper(II) atom is coordinated in the soft cavity of eob macrocycle, by two nitrogens and two phenolate oxygens. Each barium(II) atom is complexed in a perching fashion by the macrocycle. Two phenolate oxygens serve as a bridges between two metal atoms with Cu...Ba separation of 3.69 and 3.73Å. Dark brown derivative contains two crystallographically independent molecules.

In $[(en)_2(H_2O)Cu(\mu\text{-edta})Ni].4H_2O$ [66] the Ni(II) atom is surrounded octahedrally by four O and two N atoms of $edta^{4-}$ ligand. One of the coordinated carbonyl oxygen atoms is present simultaneously in the coordination of the Cu(II) atom with Cu - O - Ni bridge angle of 118.4(2)° and Cu...Ni separation of 3.844Å.

In blue $[Cu(\mu\text{-cdta})Zn(H_2O)_5].H_2O$ [67] the Cu(II) atom is surrounded octahedrally by four O and two N atoms of $adta^{4-}$ ligand. One of the coordinated oxygen atoms serve as a bridge between CuO_4N_2 and ZnO_6 moieties (Table 2).

In another blue $[(H_2O)_4Cu(\mu\text{-cdta})Ni].3H_2O$ [68] the Cu(II) atom occupies a "hydrated" octahedral site and the Ni(III) atom a "chelated" one. The two metal atoms are linked through a bridging oxygen belonging to a carboxylate group.

Red brown CuCr derivative [59] consists of a dicationic $[(H_2O)Cu(\mu\text{-dopn})\text{-}Cr(MeO)(tmac)]^{2+}$ unit, two disordered perchlorate anions, and water molecule. The copper(II) atom is in a distorted square-pyramidal environment CuN_4O and chromium(III) atom is in a distorted octahedron. These two moieties are oximato-bridged with Cu...Cr separation of 3.86Å.

There are four CuFe derivatives [69a, 71] which contain $[mnt)Cu(\mu\text{-mnt})\text{-}Fe(tpp)(thf)]^-$ anion. The heterobinuclear anion in each compound consists of an iron(III) porphyrin that is bound to a $[Cu(mnt)_2]^{2-}$ anion by coordination of one of the sulfur atoms of the copper(II) complex at an axial position, in each case, the sixth coordination site about the iron(III) atom is occupied by a thf molecule (Table 2).

Structures of two CuFe derivatives are similar [69b], both consisting of separate square-pyramidal halves, $[Fe(III)(oep)]^+$ and $Cu(bnpy_2)]^{2+}$, linked by an unsupported fluoride bridge. The bridges are not linear, the Cu - F - Fe angles being 171.9(1) and 166.2(5)°, with Cu...Fe separations of 3.956(1) and 4.047(3)Å, respectively.

Structure of green $[Cu(\mu\text{-acacP})_2PtCl_2].2CH_2Cl_2.H_2O$ shows [70] that the two acacP ligands bridge approximately square-planar Pt(II) and Cu(II) centers. The $PtCl_2$ and Cu(II) moieties are accommodated within the P_2 and O_4 compartments, respectively. The $PtCl_2P_2$ atoms are planar within 0.12Å while the $Cu(\mu\text{-diketonate})_2$ portion exhibits a greater distortion toward tetrahedral geometry with CuO_4 being coplanar with 0.21Å. The plane defined by the coordination spheres of the two metals form a 67° angle. The Cu...Pt distance of 3.966(1)Å confirms the absence of any bonding interaction.

Structure of black [(mnt)Cu(μ-mnt)Ni(tba)(MeCN)].MeCN [72a] consists of discrete binuclear CuNi unit with a unique bridging mode. The Cu(II) atom is coordinated by four sulphur atoms of the maleonitriledithioalate ligands in a distorted square-planar environment. One mnt ligand linked through one of its cyano groups to an axial site of the Ni(II) atom completing an octahedral coordination (NiN_6) (Table 2).

Structure of [Cu(μ-dpb)MnCl].C_8H_{10} [72b] is shown in Fig. 3. The Cu...Mn distance is 4.127(3)Å and both porphyrin moieties are slipped by $\alpha = 25.8°$.

In red [(Me_5dien)Cu(μ-ac)Fe(oep)]$(ClO_4)_2$.CH_2Cl_2 [73] the acetate group serve as a bridge between two (Me_5dien)Cu and Fe(oep) moieties in a *syn-anti* configuration with Cu...Fe separation of 4.430Å. The Cu(II) atom is in a square-planar environment, CuN_3O and the Fe(III) atom is in a square-pyramidal environment, FeN_4O (Table 2).

In blue-violet $(en)_2$Cu(μ-Cl)$MnCl_3$ [74a] a chlorine atom serve as a bridge between two moieties, $Cu(en)_2$ and $MnCl_3$ with Cu - Cl(bridge) distance 2.804(2)Å and Cu - Cl - Mn bridge angle of 125.3(1)° and Cu...Mn separation of 4.612(2)Å. In original paper it was proposed, that the structure consists of chains in which alternating Cu(II) and Mn(II) atoms are bridged by chlorine ligands. Seems to us, that the second Cu - Cl(bridge) distance of 3.063(2)Å is too long for "real" bond.

There are two dark violet CuFe derivatives [74b] in which cyanide group serve as a bridge in the manner of Cu - NC - Fe with Cu(II)...Fe(III) separation of 4.94 and 4.98 Å, respectively (Table 2).

Structure of dark violet Cu(μ-Cl_4obbz)Mn$(H_2O)_5$ [75] consists of binuclear units with Cu(II) and Mn(II) atoms bridged by a carboxylate group with the *anti-syn* configuration with Cu...Mn separation of 5.195(1)Å. The copper(II) atom is in a square-planar environment CuO_2N_2 and the Mn(II) atom is in a pseudo-octahedral, MnO_6.

Structure of red-brown CuNi derivative [76a] consists of isolated heterobinuclear units with square-planar Cu(II) and octahedral Ni(II) atoms bridged by an oxamido group with Cu...Ni separation of 5.235(1)Å (Table 2).

Structure of blue red [Cu(μ-oxap) Ni(phen)$_2$](ClO$_4$)$_2$.2H$_2$O [76b] is similar to that of CuNi derivative [76a]. Two moieties a square-planar CuN$_4$ and an octahedral NiN$_4$O$_2$ are bridged by an oxamido group with Cu...Ni separation of 5.292 Å.

Another three heterobinuclear CuNi derivatives [77, 78] have a similar structures to that of the earlier examples [76] with Cu...Ni separations of 5.29 and 5.32 Å [77] and 5.338 Å (Table 2).

Structure of violet derivative [79] consists of oxamido-bridged Cu^{II}Mn^{II} discrete units [Cu(μ-obze)Mn(H$_2$O)$_4$] and noncoordinated water molecules. The copper(II) atom is coordinated to the two oxamido nitrogen atoms and two carboxylato oxygen atoms in a slightly distorted square-planar arrangement. The Mn(II) atom has somewhat distorted octahedral surroundings with four water molecules and the two oxamido oxygen atoms in the coordination sphere. The Cu...Mn separation of 5.422(1)Å ruled out direct metal-metal bond.

In another red Cu^{II}Mn^{II} derivative [80] oxamido-bridge held together a square-planar CuO$_4$ and a pseudo-octahedral MnN$_4$O$_2$ moieties within Cu...Mn separation of 5.436Å.

Dark derivative [81] has an oxalate-bridged binuclear Cu(II) - Cr(III) core with a Cu...Cr separation of 5.482(3)Å. The geometry around the Cu(II) atom is square-pyramidal with the three donor atoms of acpy and one of the oxalate oxygens at the basal plane and the other oxygen of the oxalate group at the apex.

In blue CuZn derivative [82], two equivalent moieties, Cu(tren) and Zn(tren) are held together by an imidazolate ligand with Cu...Zn separation of 5.84Å. Both metal atoms are surrounded by five N donor atoms and have an approximately trigonal-bipyramidal geometry.

In violet CuCo derivative [83] the imidazolate ligand bridged two units, (ClO$_4$)(pmdt)Cu and Cu(NH$_3$)$_5$, with Cu(II)...Co(III) separation of 5.859(2)Å. The Cu(II) atom is in a square-pyramidal (CuN$_4$O) and the Co(III) atom in an octahedral (CoN$_6$) environment.

In green CuZn derivative [84], two trigonal bipyramids, CuN_4O and ZnN_4O, are connected by a phosphate group of p-$O_2NC_6H_4OPO_3$ ligand, with Cu...Zn distance of 5.86 Å.

Structure of $[Cu(\mu\text{-im})(\mu\text{-cryptand})Zn]^{3+}$ [85] is shown in Fig. 4. Both metal(II) atoms are surrounded by five N atoms, the Cu(II) atom in a slightly distorted trigonal-bipyramidal and Zn(II) atom in a distorted square-pyramidal environment, with Cu...Zn separation of 5.93Å (Table 2).

X-ray analysis of blue CuZn derivative shows [87] that the derivative consists of a $[(trien)Cu(\mu\text{-im})Zn(trien)]^{3+}$ cation, three ClO_4^- anions, and one water molecule. In the cation, the Cu(II) and Zn(II) atoms are bridged by the imidazolate anion, forming a heterobinuclear complex with Cu...Zn separation of 6.081(2)Å. Both metal atoms, MN_5, are in a distorted square-pyramidal environment.

X-ray analysis of red brown CuMn derivative confirmed [88] an imidazolate bridged copper(II)-manganese(III) binuclear structure, in which the coordination geometries of the copper(II) and manganese(III) atoms are square-planar (CuN_3O) and square-pyramid (MnN_3O_2).

In purple brown CuCo derivative [89], a pyrazinate bridged copper(II)-cobalt (III) binuclear structure, in which the coordination geometries of copper(II) and cobalt(III) atoms are square-planar (CuN_4) and an octahedral (CoN_6).

In violet CuNi derivative [90] are an extanded oxamido-bridged a distorted octahedral (CuN_4O_2) and a square-planar (NiN_4) units. In red brown CuPd derivative [91] the Pd-C σ-bonded 1,4-diaza-3-methylbutadien-2-yl group acts as σ,σ-N,N-chelating ligand with a tetragonally distorted tetrahedral geometry arround the Cu(II) atom, CuN_2Cl_2.

In brown CuPd derivative [92] two moieties, $CuCl_2$ and $PdBr_2$ are connected by pentadentate-N_5-hexakis(3-pyrazolyl)cyclotriphosphazene via simultaneous nongeminal (N_3) to Cu(II) and geminal (N_2) to Pd(II) atoms. The copper(II) atom as a trigonal-bipyramidal (CuN_3Cl_2) and palladium(II) atom a square-planar (PdN_2Br_2) environment.

The asymmetric unit of purple $Cu\{\mu\text{-(dpe)py}_2\}PtCl_2$ [93] contains two crystallographically independent molecules. Each metal(II) atom has a square-planar geometry, CuN_4 and PtN_2Cl_2 (Table 2). In another purple CuZn derivative [94] two moieties, $Cu(C_{10}H_{24}N_4)$ and $ZnCl_3$ are bridged by a single chlorine atom with Cu - Cl

- Zn bridge angle of 113.14(2)°. The copper(II) atom is in a square-pyramidal (CuN_4Cl) and zinc(II) atom in a tetrahedral ($ZnCl_4$) environment. Two non-equivalent units, CuN_5O and NiC_4, are bridged by cyanide group in the manner, Cu-NC-Ni. In blue CuV derivative [96] an oxalate group serve as a bridge between two heteroatoms. The copper (II) atom is in a trigonal-bipyramidal (CuN_3O_2) and vanadium (IV) atom in a pseudo-octahedral (VO_6) environment.

There are almost one hundred heterobinuclear copper(II) derivatives. There are several types of bridges, of which a double bridge of the Cu - $(OL)_2$ - Cu type is the most common. In general, in the molecular entities, the copper(II) occupies the O_2N_2 inside site, whereas the other metal occupies the O_2O_2 outside site.

Inspection of the data in Table 2 reveals, that the most common mono- and multidentate ligands are O- and N- donors. There are uni- bi-, tri-, tetra-, penta-, hexa-, hepta- and octadentate ligands. The data in Table 2 indicates copper(II) in four- (mostly squar-planar), five- (mostly square-pyramidal) and six- (mostly tetragonal-bipyramidal) coordination environments, from which the two former prevail. The mean Cu - L bond dostance in the series of four- coordinated derivatives increases in the sequence: 1.94Å (O_2L) < 1.985Å (N_6L) < 1.99Å(N_2L; N_4L) < 2.00Å (N_3L) < 2.17Å (Cl) < 2.24Å (OL) < 2.29Å (S_2L). For the five- coordinated species the order is: 1.98Å (N_2L) < 2.03Å (N_3L) < 2.07Å(N_4L) < 2.09Å (O_2L) < 2.12Å (OL) < 2.13Å (N_3L) < 2.14Å (NL) < 2.22Å (O_4L) and for the six- coordinated species the order is: 2.00Å (N_4L) < 2.03Å (N_2L) < 2.12Å (O_2L) < 2.13Å (NL) < 2.32Å (OL).

The multidentate ligands include both homo- and hetero- donor atoms. The multidentate ligands are: bidentate with two O, two N and two S atom donors; tridentate with three N, and one O plus two N atom donors; tetradentate with four O, four N, three O plus N; two O plus two N, and one O plus three N atom donors; pentadentate with one O plus four N atom donors; hexadentate with six N, four O plus two N; four O plus two P, and two O plus four N atom donors; heptadentate with five O plus two N, and one O plus six N atom donors; and octadentate with eight N, six O plus two N and two O plus six N atom donors sites.

Both steric and electronic factors associated with the donor atom appear to influence the L - Cu - L bond angles of the various metallocyclic rings. The mean values for the angles are: five-membered rings: 82.3° (O+N-donor) < 85.7° (N-donor) < 91.0° (S-donor) for four- coordinated species; 76.8° (O-donor) < 82.0° (O+N-donor) for five- coordinated species, and 82.0° (O-donor) < 82.5° (O+N-donor) < 83.0° (N-donor) for six- coordinated species; six- membered rings: 94.3° (N-donor) < 94.5° (O-donor) < 94.7° (O+N-donor) for five- coordinated species, and 91.5° (O-donor) < 93.5° (O+N-donor) < 95.0° (N-donor) for six- coordinated species.

The (MeOH)Cu{μ-$(fsa)_2$en}VO [36] exists in two isomeric froms, red and dark blue, which differ mostly by degree of distortion in both the M - L bond distances and L - M - L angles. Two independent molecules differing by degree of distortion have been found [(ClO_4)Cu(μ-nup)Ni(Me_3[12]N_3)]$ClO_4.H_2O$ [57a], Cu(μ-eob)Ba$(CF_3SO_3)_2$ [65] and Cu{μ-(dpe)py_2}$PtCl_2$ [93]. The coexistence of two or more species differing only by degree of distortion is typical of the general class of distortion isomerism [97].

The Cu - M distances are summarized in Table 2A, where the variety of metals which can interact with the bordeline copper(II) atom can be seen. The metal atoms cover the range from main group to actinide. This illustrates the rich diversity of the chemistry of copper(II). It should be noted that there are several examples for which full data have not been published. The Cu(II) - M (non-transition distances vary from 2.778(4) (M = Al(III)) [38a] to 6.281(2) (M = Zn(II)) [87]; for M = transition metal distances vary from 2.558(1) (Mo) [31] to 5.859(2) (Co(III)) [83], and for M = lanthanoide, actinoide, in the range from 3.355(3) (Pr(III)) [53] to 3.538(2)Å (Pr(III)) [58].

There are some examples, in which Cu - M distance elongated with an opening Cu - X - M bridge angle. For example, for M = Zn(II) the values are: 2.966Å vs 93.8° [37] < 2.985Å vs 100.3(2)° [39] < 3.186(2)Å vs 102.4(3)° [50]; for M = Pr (III) the values are: 3.355(3)° vs 101.2(5)° [53] < 3.538(2)Å vs 105.5(3)° [58]; for M = V(IV) : 2.949(3)Å vs 95.1(4)° [36] < 2.989(4)Å vs 98.6(7)° [36]; for M = Fe(III) : 3.20Å vs 100.0(4)° [42] < 3.401(4)Å vs 119.3(5)° [54] < 3.641(1)Å vs 122.5(2)° [54]; for M = Co(II); 3.025(2)Å vs 99.1(2)° [43] < 3.03Å vs 99.9(5)° [42] < 3.060(1)Å vs 99.8(2)° [47]; and for M = Ni(II): 2.897(2)Å vs 93.9(1)° [33b] < 2.975Å vs 98.9(1)° [38] < 3.038(2)Å vs 99.9(3)° [46] < 3.844Å vs 118.4° [66].

4. HETEROTRINUCLEAR COPPER (II) COMPOUNDS

The crystallographic and structural data for these sixty derivatives are summarised in Table 3. They are tabulated in order of decreasing number of copper(II) atoms. There are over twenty derivatives which contain two copper(II) atoms and the third one is non- transition, transition or lanthanide, and remaining almost forty species contain one copper(II) atom and remaining two are non-transition, transition or lanthanide. Heterotrinuclear copper (II) derivatives are with variety of types of bridging.

In nine Cu_2M derivatives [98-104] "real" central atom is M and two CuL complex ligands bind to a central M by two O double bridged ligands. In these derivatives Cu - M distance range from 2.976Å (M = Te(II)) [98] to 5.34(5)Å (M = Gd(III)) [101] and Cu...Cu distance from 3.5Å [104] to 5.402(3)Å [101]. Copper(II) atoms are found in a square-planar environment [98,100,103,104], a trigonal-bipyramidal [99], and square-pyramidal [101]. There are two derivatives [101,102] which contain non-equivalent copper(II) units, one a square-planar and the other one a square-pyramidal. Coordination about "real" central atom M are a pseudo-octahedral [98,99], a pentagonal-bipyramidal [101,102], a square antiprism [100,103]. a tricapped trigonal prism [102] and a distorted icosahedral geometry [104] (Table 3).

Structure of yellow-green Cu_2Zn derivative [105] consists of linear Cu(II)-Zn(II) - Cu(II) trinuclear unit which has an inversion centre. The Cu(II) and Zn(II) are bridged by the two phenolate oxygens of the salpd and the oxygens of the acetate ion. The coordination geometry of the terminal copper(II) atoms is square-pyramidal (CuO_3N_2) and about zinc(II) atom is a distorted octahedral (ZnO_6). The Cu...Zn separation is 3.038(2)Å and mean Cu - O - Zn bridge angles are 96.2(4)°.

Two purple Cu_2M (M = Ca or Sr) derivatives [106] have similar structures. They are composed of a three-sided Cu - Cu - M core, where each metal(II) atom is chelated by a ß-diketonate ligand. One alkoxide group bridges the heterometal and one Cu(II), whereas the other two alkoxides link all three metals. The alkoxide moieties are all folded back such that their internal O-atoms get more, strongly exposed to be heterometal. The Cu...Ca separation of 3.152(2) and 3.318(2)Å are somewhat shorter than the Cu...Sr separation of 3.259(2) and 3.394(2)Å.

Structure of dark blue Cu_2Y derivatives [107] is shown in Fig. 5. The ethoxy O atoms of the 2-methoxyethoxides (moe) connect the Cu(II) and Y(III) centers, where by two (O(9), O(11) and one μ_3 bridge (O(13)) exist. The methoxy O atoms of the μ-bridging ligands are attached to the Y center. The Cu - Y distances are 3.265(2) and 3.286(2)Å and Cu - Cu distance is 3.071(2)Å and thus beyond the limits of a covalent M - M contait.

There are two blue Cu_2M (M = Nd or La) derivatives [53,108] which have similar structures. The three metalatoms arranged in a triangular geometry with a Cu...Cu separation of 3.165(5)Å and Cu...Nb of 3.536(3)Å [53] and Cu...Cu separation of 3.265(6) and Cu...La of 3.632(4)Å [108].

In violet Cu_2Fe derivative [73] a "real" central Fe(II) atom is bound to two terminal (Me_6tren)Cu units by μ-HCOO groups from top and bottom, with Fe...Cu separation of 5.422Å. Each Cu(II) atom has a distorted trigonal-bipyramidal environment (CuN_4O), and Fe(III) atom a pseudo-octahedral (FeN_4O_2).

Another two heterotrinuclear derivatives dark green Cu_2Cr and dark blue Cu_2Mo [109] have similar structures as well. The two Cu(acpa) moieties are bridged in the manner of (acpa)Cu-O-M(O_2)-O-Cu(acpa) (M = Cr or Mo), with Cu...M separation of 3.3961(9)Å (M = Cr) and 3.576(1)Å (M = Mo) and Cu - O - M bridge angles of 141.3(2) and 149.0(4)°, respectively. the Cu...Cu separations are 6.443(1) and 6.703(2)Å, respectively. Each copper (II) atom has a square-planar (CuO_2N_2) and M(III) atom a tetrahedral (MO_4) envinonment.

In dark violet Cu_2Fe [74b] and dark blue Cu_2Fe [110] derivatives a cyanide groups serves as bridges in the manner of Cu-NC-Fe-CN-Cu (Table 3). In purple brown Cu_2Co derivative [111] the imidazole ligands connect one Co(III) and two Cu(II) atoms, forming an imidazolate-bridged heterotrinuclear complex. The Cu...Co and Cu...Cu separations are 5.853 and 11.705Å.

In dark green Cu_2Zn derivative [112] all three metal(II) atoms are "incorporated" into the 38-membered macrocycle having eight amine and four phenol functionalites. The $[Cu_2(\mu\text{-tem})Zn]^{3+}$ cation has a helically twisted configuration. The structure consists of coordination polyhedra around heterobinuclear CuZn core and isolated copper site. The Cu...Zn separations are 3.08 and 5.61Å and Cu...Cu separation is 8.13Å (Table 3).

In ruby violet derivatives, Cu_2Ni [113] and Cu_2Gd [114] two oxamidate anions act as bridges between the central Ni [113] or Gd [114] and the outer copper(II) atoms. Each central atom is coordinated by two pairs of oxamidate oxygens and two water molecules in [113] completed an octahedral environment about Ni(II), and four water molecules [114] completed eight- coordination about Gd(III) atom. The oxamidate anions are functionalized by two N-bonded three-carbon chains dimethyl substituted at the central carbon and terminating with amino groups. The resulting N-tetradentate ligands coordinate in a planar geometry two copper(II) atoms. The mean Cu...M distances are 5.316(1)Å (M = Ni) and 5.683(1)Å (M = Gd), respectively.

Almost forty CuM_2 heterotrinuclear derivatives show a rich variety of bridges types.

There are five CuM_2 derivatives (M = Co(III)) [115] and (M = Cr(III)) [116] in which the central copper(II) atom is bonded to the oxygen atoms of four bridging hydroxo groups. The hydroxo groups bridge, in pairs, to two cobalt or two chromium atoms. While each M(III) atom is a pseudo-octahedrally coordinated (MN_4O_2); the copper(II) atoms are four-, five- and six-coordinated (Table 3). The mean Cu - Co distances are 2.888(1) and 2.899(1)Å [115], and the mean Cu - Cr distances are 2.899(2) and 2.958(1)Å [116].

In deep red $[(OC)_4Re(\mu\text{-}\eta^2\text{-}MeCO)_2Cu(\mu\text{-}\eta^2\text{-}MeCO)_2Re(CO)_4]$ [117] the central copper(II) atom is coordinated to two metallaacetylacetonate ligands where all four Cu - O distances are 1.920(3) Å. Although the CuO_4 coordination geometry is rigorously planar, the atoms comprising the backbone of each metallaacetylacetonate ligand are not coplanar, resulting in a unique "chaise longue chair" structure for the molecule which preserves the crystallographically imposed inversion symmetry. The Cu...Re separation is 3.617(0) Å.

The lilac $[\{Me_2Ga(\mu\text{-}pz)_2\}_2Cu]$ molecule [118] is in a pseudo-chair corformation with the two six- membered Ga - $(N\text{-}N)_2$ - Cu rings in boat conformations. The copper(II) atom lies on a crystallographic centre of symmetry in the middle of a planar arrangement of four nitrogen atoms. Steric requirements of the ligands in $[\{Me_2Ga(\mu\text{-}Mepz)_2\}_2Cu]$ [118] necessiate adaption of pseudo-tetrahedral geometry about the central copper(II) atom, with dihedral angle 71.9° between the two N - Cu - N coordination planes, and an almost planar arrangement for the Ga - $(N\text{-}N)_2$

- Cu six-membered rings. There are six $CuPt_2$ derivatives, two orange-red [119] and remaining four are green of colour [120-123], in which metal(II) atoms are doubly bridged by three-atoms bridges in the manner of Pt-(N-C-N)$_2$-Cu-(O-C-N)$_2$-Pt. In each of the derivatives, the central copper(II) atom lies on a crystallographic centre of symmetry, in the middle of a planar arrangement of four oxygen atoms. The bridging ligands are 2-pyridonate [119], 1-methyluracilate [120,121,123] or acetamidate [122].The Cu - Pt distances range from 2.6312(4)Å [119] to 2.9843(1)Å [123]. In all derivatives, each metal(II) atom is in a square-planar arrangement (CuO_4 and PtN_4) (Table 3).

In $CuRu_2$ derivative [124] three metal atoms are triply bridged. The centrosymmetric structure consists of a copper(II) atom which is tetragonally coordinated by four oxygen and two chlorine atoms (CuO_4Cl_2). The chlorine atoms bridge directly between the copper(II) and ruthenium(II) with Cu - Cl - Ru bridge angle of 93.6(1)$^\circ$. The other bridging ligands are the $OP(OMe)_2$ ligands, which coordinate to copper with oxygen and to ruthenium with phosphorus. The ruthenium coordination is completed by the η^6-C_6Me_6 ligand.

In $CuFe_2$ derivative [125] the $Fe(paba)_3^-$ fragment act as a facially dispose tridentate oximato - O_3 ligand and the motif Fe-(N-O)$_3$-Cu-(O-N)$_3$-Fe is present. The nonbonded Cu...Fe distance is 3.483(1)Å and Fe...Cu...Fe fragment is exactly linear. The CuO_6 coordination sphere is strongly Jahn-Teller elongated tetragonal bipyramid (Table 3).

Structure of dark $CuMn_2$ derivative [126] consists of distinct $[(tmac)_2Mn_2(\mu$-niox$)_3Cu]^{4+}$ cations, perchlorate anions and acetonitrile molecules. The central Cu(II) atom is six- coordinated by azomethine nitrogen atoms of the three nioxime ligand. The $Cu(niox)_3^{4-}$ anion bridges two terminal Mn(IV) atoms through its deprotonated oxime oxygens in the motif of Mn-(O-N)$_3$-Cu-(N-O)$_3$-Mn with a Mn...Cu separation of 3.54Å. The trinuclear Mn...Cu...Mn unit is not perfectly linear with an angle of 171° and the Mn...Mn separation is 7.06Å.

In $CuFe_2$ derivative [127] the $Cu(dmg)_3^{4-}$ anion bridges the two Fe(III) atoms through its deprotonaled oxygens in the matter of Fe-(O-N)$_3$-Cu-(N-O)$_3$-Fe with a Cu...Fe separation of 3.58Å and Fe...Fe separation of 7.0Å. Two facially coordinate tridentate amine Me_3tac ligands complete the trigonally distorted octahedral coordination sphere of the two Fe(III) centres (FeO_3N_3) (Table 3).

In brown $CuLi_2$ derivative [128] the $Cu(tdp)_2^{2-}$ anion acts as a bridging ligand between the Li(I) atoms using the oxygen atoms of its thiodiphenolate ligands, with the mean Cu - O - Li bridge angles of 91.4°. The coordination sphere of each Li(I) atom is completed by two methanol ligands which leads to distorted tetrahedral coordination. The Cu(II) atom has approximately octahedral geometry (CuO_4S_2). The tridentate 2,2′-thiodiphenolates occupy facial positions.

There are two green CuM_2 (M = Ti(IV) [129] and Zr(IV) [130]) derivatives which are isostructural. The molecule is an isosceles triangle of metal atoms with non-bonded distances (Table 3). The triangle is held together by three μ- and two μ_3-O^iPr groups. The chloride is found exclusively as a terminal ligand on Cu(II) atom. Thus, the stronger π-donor O^iPr groups remain attached to the stronger Lewis acid M(IV). Formally this molecule may be viewed as a $CuCl^+$ fragment which has a $M_2(O^iPr)_9^-$ unit coordinated to it in tetradentate fashion.

In turquoise blue derivative [131] the $CuBa_2$ framework is based on a triangular unit capped by two triply bridging aminoalkoxide ligands. The most salient feature is the disymmetry of the molecule. The Cu(II) atom is in a trigonal-bipyramidal environment (CuO_4N). Barium(II) atoms are differ from each other, one is eight coordinated (BaO_8) and the other is seven coordinated with a copper octahedral surrounding (BaO_6N). The Cu...Ba separations of 3.325(2) and 3.520(2)Å as well as Ba...Ba separation of 4.074(2)Å are non bonding.

Structure of another derivative [132] consists of $[Cu(en)_2(H_2O)_2]^{2+}$ cations, $[\{Ni(\mu\text{-}edta)\}_2Cu(en)_2]^{2-}$ trinuclear anions, and water molecules. In the anion, the Cu(II) atom lies at a center of symmetry and is bonded to two ethylenediamine ligands, each Ni(II) atom is octahedrally coordinated by the hexadentate edta ligand; one of its carboxylate O atoms bonded to Ni expands the square coordination of Cu to octahedral. The bridging oxygens join the apices of the octahedra into a trinuclear associate, with Cu - O - Ni bridge angle of 117.4(1)° and Cu...Ni separation of 3.788Å.

Structure of brown $CuMn_2$ derivative [133] is shown in Fig. 6. This structure consists of centrosymmetrical M(III) - Cu(II) - Mn(III) trinuclear units and non-coordinated dmso molecules. Each metal is six- coordinate with the chromophores CuN_4O_2 and MnO_4N_2. The Cu...Mn and Mn...Mn separations of 3.836(1) and 7.671(2) Å ruled out direct metal-metal bond.

In deep blue derivative [134] two niobium(V) atoms and a copper(II) atom are linked by two oxo bridges, Nb - O - Cu - O - Nb. In another deep blue $CuNb_2$ derivative [134] the metal are bridges in the manner of Nb - O - Cu - F - Nb. In both derivatives, each metal atom is a pseudo-octahedrally coordinated.

In $CuCr_2$ derivative [135] two Cr(VI) atoms and a Cu(II) atom are linked by two oxo bridges, but in the manner of Cu-O-Cr-O-Cr. The Cu(II) atom has a trigonal-bipyramidal environment (CuN_4O) and both Cr(VI) atoms are tetrahedrally coordinated (CrO_4).

In brown $[(NH_3)_5Co(\mu\text{-im})Cu(dien)(\mu\text{-im})Co(NH_3)_5]Cl_6$ [136] the imidazole ligands serve as bridges between metal atoms. The Cu(II) atom geometry (CuN_5) is intermediate between trigonal bipyramid and square pyramid. Each Co(III) atom is six- coordinated (CoN_6).

X-ray analysis of dark brown $\{(CO)_5Mo(\mu\text{-dps})\}_2Cu$ shown [137] that there are two crystallographically independent molecules. The molecules differ primarily in their conformation about Mo - P bond as described by the C - Mo - P - N torsional angles of 70.7(1) and 158.4(1)^o for the respective molecules. The Cu(II) atom has a square-planar geometry (CuO_2N_2) and Mo(0) atoms are six-coordinated, MoC_5P. The separations between the central Cu(II) and terminal Mo(0) atoms are 7.078(1)Å for one and 7.079(1)Å for the other molecule.

Thre are five $CuFe_2$ derivative [138,139] of the composition $[\{(cp)Fe(\mu\text{-}C_5H_4CO_2)_2\}_2CuL_2]$ in which two terminal ferrocene-carboxylate ligands are bound through O-atoms to the central CuL_2 unit (L = 1,2 dimethylimidazole [138], N-methylimidazole [138], pyridine [139] or imidazole [139]).

Structure of dark brown $\{(cp)Fe(\mu\text{-hbcpa})\}_2Cu$ shows [140] that each of the two terminal N-(O-hydroxybenzylidene) ferroceneamine ligands through O- and N-donor atoms bonded to central Cu(II) atom in a distorted environment. The Cu...Fe separations of 4.300(1) and 4.350(1)Å, respectively, thus excluding any significant direct interaction.

Finally, in green $\{(CO)_5W(\mu\text{-odp})\}_2Cu$ derivative [140] each of the two $(CO)_5W(odp)$ terminal ligands is coordinated to the central Cu(II) atom through two O-donor atoms.

Inspection of the data in Table 3 reveals that there are homo- and heterodonor ligands; uni-, bi-, tri-, tetra-, hexa-, octa-, undeca- and even tetradecadentate ligands.

The data indicates copper(II) in four- (mostly square-planar), five- (mostly square-pyramidal) and six- (mostly tetragonal-bipyramidal) coordination environments, from which the two former prevail. The mean Cu - L bond distance in the series of four-coordinated derivatives increase in the sequence: 1.92Å (μ-O) < 1.935Å (μ-OH) < 1.97Å (N_2L) < 1.98Å (NL); for the five-coordinated derivatives the order is: 2.00Å (N_2L) < 2.02Å (μ-OL) < 2.10Å (N_3L) < 2.11Å (N_4L) < 2.17Å (Cl) < 2.50Å (OL).

In the series of heterodonor ligands are bidentate: O + N and O + S; tridentate: O + 2N; tetradentate: 2O + 2N; hexadentate: 4O + 2N; octadentate: 2O + 6N; unidecadentate: 4O + 7N and tetradecadentate: 10O + 4N donor atom sites.

Both steric and electronic factors associated with the donor of the various metallocyclic rings. The mean values for the angles are: five-membered rings, 82.5° (O-donor) < 85.0° (N-donor) ~ 85.0° (O+N-donor); six- membered rangs, 93.5° (O-donor) ~ 93.5° (O+N-donor) < 94.2° (N-donor).

A summary of the metal-metal distances is given in Table 3A where the variety of metals which can interact with the borderline copper (II) atom can be seen. In the series of Cu(II) with non-transition metals, the Cu...M separation range from 3.038(2)Å (M = Zn(II) [105] to 5.61Å (Zn(II)) [112]; in the series of Cu(II) with transition metals in the range of 2.6312(4)Å (M = Pt(II)) [119] to 7.079(1)Å (Mo(II)) [137], and in the series of Cu(II) with lanthanides the range is 3.30(1) (M = Gd(III)) [101] to 5.684(1)Å (Nd(III)) [114]. However it must be noted, that there are papers in which metat-metal separations are not given.

In $\{(CO)_5Mo(\mu\text{-dps})\}_2Cu$ [137], two crystallgraphically independent molecules are present which differ by degree of distortion and it is another example of distortion isomerism [97].

5. HETEROTETRANUCLEAR COPPER(II) COMPOUNDS

The data for these compounds are assembled in Table 4. The structures are tabulated in order of decreasing number of copper(II) atoms. There are seven complexes of heterotetranuclear compounds which contain three copper(II) atoms. In the structure of bright green tris-copper-magnesium derivative [142] six 2-hydroxy-6-methylpyridine anions serve as bridges by O- and N- donor atoms between the metal(II) atoms. Each Cu(II) atom has an almost square-planar arrangement (CuO_2N_2) and Mg(II) atom is six-coordinated (MgO_6).

Structure of dark blue Cu_3Y derivative [107] consists of a four-metal atom core having Cu-Y distance of 3.204(2), 3.246(2) and 3.377(2)Å and Cu...Cu contacts of 3.107(2) and 3.711(2)Å. Each of metal atoms is bonded to the O atoms of one β-diketonate ligand. The linkage of the metal cores is established by five 2-methoxyethoxide units . The alkoxide O atoms of the three 2-methoxyethoxide ligands are function as μ- and remaining two as μ_3-bridges. The Y(II) center is surrounded by eight O atoms and each Cu(II) atom has a square-pyramidal geometry (CuO_5).

Structure of green Cu_3La derivative [143] is a highly asymmetric, with each of the three Cu(II) atoms in chemically distinct sites. Atom Cu(1) is four coordinate, bound to two nitrogens from 6-chloro-2-hydroxypyridine (chp) rings, one μ-oxygen from a third chp, and one μ-oxygen from a deprotonated methanol. These two μ-O atoms are shared with Cu(3). Atom Cu(1) therefore has a *cis*-O_2N_2 array of donor atoms. Atom Cu(II) also has a *cis*-O_2N_2 set of donor atoms, with all the donors coming from chp units. The oxygen atoms are again μ-bridging, this time between Cu(2) and La. Atom Cu(3) has a quite different coordination sphere CuO_4N. Two of the oxygens are the pair shared with Cu(1), one is from a chp unit, and the final oxygen comes from a terminal molecule of methanol. The nitrogen bound to Cu(3) comes from a chp unit. The La(III) atom is ten coordinate, bound to six oxygens of bidentate nitrate anions, two μ-oxygens from chp units shared with Cu(2), one further oxygen from a chp ligand, and finally a molecule of MeOH (Table 4).

In another green Cu_3La derivative [143] also all three copper sites are chemically distinct, although this time all are four coordinate, close to square planar geometries: atom Cu(1) is bound to three nitrogens and one oxygen from chp units. Atom Cu(2) has a *cis*-arrangement of 2 O and 2 N atoms from chp units, the two oxygens being μ-bridging of La. Atom Cu(3) is bound to 2 O and 2 N with a *trans*-geometry, with two nitrogens and one of the oxygens coming from chp units, and the final oxygen from a molecule of EtOH. The La(III) atom is eight-coordinate bicapped trigonal prism, bound to two μ-oxygens shared with Cu(2), four mononucleating oxygens from chp units, and a bidentate nitrate ion.

There are three dark green Cu_3M (M = Gd, Dy or Er) derivatives [144] which are isostructural. Structure of $Cu_3Gd(chp)_8(NO_3)H_2O)$ is shown in Fig. 7. The four metals are arranged in an approximately planar quadrilateral. The shortest contact is that between Cu(1) and Gd(1) (3.627Å), which is bridged by two μ-O atoms, while the Cu...Cu contacts are between 3.61 and 3.70Å.

In another twenty five examples there are two copper(II) atoms and two other metal atoms [48,51,58,106,107,143,145-157] (Table 4). Structure of green $[\{Cu(\mu\text{-}bdd)K\}(\mu\text{-}pcr)]_2.2CHCl_3$ [145] consists of a centrosymmetric dimer. The Cu(II) atom is cooordinated intramolecularly by two β-diketone units located at both of the terminals of the same molecule in a nearly planar tetradentate structure (CuO_4). By this coordination, four neutral ether oxygens and two β-diketonate oxygen atoms are assembled to constitute a pseudo-crown ring, in which the K(I) atom is entrapped. The K(I) atom is not fully buried in this pseudo-crown ring but pulled out of the plane toward two picrate anions with asymmetric bond lengths of 2.703(4) and 2.742(4)Å. Two K(I) atoms are bridged by two picrate anions to make a "sandwich dimer" structure of 2:2 stoichiometry.

In green Cu_2Rb_2 derivative [146] chlorine atoms serve as bridges between the metal atoms, (18-crown-6)Rb-$(Cl)_2$-Cu-$(Cl)_2$-Cu-$(Cl)_2$-Rb(18-crown-6), With Cu...Cu and Cu...Rb separations of 3.386(1) and 3.917(1)Å, respectively. The Rb(I) atoms are in typical "sunrise" coordination to 18-crown-6 (0.81Å above the best plane of the six crown oxygens).

Crystal structure of blue Cu_2Ca_2 derivative [147] comprises discrete $[\{CuCa(Et_2NCH_2CO_2)_4(NO_3)_2(H_2O)\}_2]^{4+}$ cations, nitrate anions and lattice water molecules. The tetranuclear cation is a dimer of heterobinuclear subunits in which a pair of Cu(II) and Ca(II) atoms separated by 3.420(2)Å are linked by a conventional tetrakis-carboxylato bridge. The Cu(II) atom is square-pyramidally coordinated by four carboxylate oxygen atoms at the basal plane, plus one oxygen atom from a carboxylate group of the other subunit at the apical position. A pair of such μ-1,1-O bridges, related by two-fold axial symmetry, connect two heterobinuclear subunits into a tetranuclear cation, constituting a Cu_2O_2 rhomb.

Dark blue Cu_2Sr_2 derivative [51] contains two binuclear $Cu(bdmap)(F_3ac)_2Ca$ units, related by an inversion center. Two binuclear units are linked together through two trifluoroacetate bridges with asymmetrical Sr-O bond distances of 2,47(1) and 2.60(1)Å. The Cu - Sr separation is 3.390(3)Å. The Cu(II) atom is coordinated by two oxygen and two nitrogen atoms of bdmap ligands in approximate square-planar geometry. The fifth position of the Cu(II) atom is occupied by an oxygen atom of trifluoroacetate ligand. The strontium(II) atom is surrounded by five oxygen and two nitrogen atoms in an approximate pentagonal-bipyramidal geometry.

X-ray analysis of dark blue Cu_2Ba_2 derivative shows [148] that the metals are arranged in a rhomboid with triply bridging alkoxide oxygens above and below the two triangular planes defined by two Ba(II) atoms and one Cu(II) atom. Each metal atom is also bound to an acac ligand, and the barium is further coordinated by the methoxy oxygen of the triply bridging alkoxides and a 2-methoxyethanol (Table 4).

Light blue derivative[149] consists of two copper(II) and two barium(II) atoms in a rhombic arrangement with the Cu...Ba distances being 3.5817(6) and 3.6251(6)Å, respectively, and the Ba - Cu - Ba and Cu - Ba - Cu angles being 64.89(1) and 115.11(1)°, respectively. The molecule of the derivative has an inversion center and contains four 2,2'-thiodiethanolate ligands. The four alkoxo oxygen atoms of the four ligands function as triply bridging atoms to the metal centers. The Ba(II) atom is surrounded by eight oxygen atoms, four alkoxo oxygen atoms and four from the two hexafluoroacetylacetonate ligands. Each Cu(II) center is chelated by two $Htde^-$ ligands via the alkoxo oxygen and sulfur atoms and created a distorted elongated octahedron about each Cu(II) atom (CuO_4S_2).

There are two purple Cu_2Ba_2 derivatives [106] which are structurally very similar to each other. Both molecules consists of a four-side Cu_2Ba_2 and are of alternating Cu(II) and Ba(II) atoms. Both molecules occupy crystallographic centers of inversion, hence each molecule exhibits only two unique Cu - Ba distances and the Cu_2Ba_2 units are planar. Each metal is bonded to one β-diketonate ligand, the Ba(β-diketonate) rings are coplanar with the Cu_2Ba_2 units, while the Cu(β-diketonate) moieties are nearly perpendicular with interplanar angles of 88.2 and 84,5 °, respectively. Each compound contains four alkoxide functions, each of which triply bridges both barium atoms and the copper center (Table 4).

Green crystal of Cu_2Ba_2 derivative [150] comprises binuclear centrosymmetric molecules $[Cu_2(dmae)_4Pb_2I_4(dmso)_2]$ bridged by oxygen atoms of chelated 2-

dimethylaminoethoxo groups, and dmso molecules of crystallization. The bridging oxygen atoms arrange a strictly planar central Pb_2O_2 rhombus. The copper coordination environment has a somewhat distorted square-planar geometry (CuO_2N_2). The coordination geometry around the Pb(II) atom is a higly distorted octahedron (PbO_4I_2). The Cu...Pb distances are 3.394(2) and 3.634(2)Å.

In blue Cu_2Ga_2 derivative [131 two {$Me_2Ga(\mu$-dmae)(μ-pz)Cu} fragments are bridged by two pyrazolyl groups. The molecular framework consists of a system of nine fused rings, five of which are crystallographically unique. The central Cu_2N_4 ring, situated about the molecular inversion center, is significantly non-planar and has a flattened chair conformation. The Cu...Cu distance across this ring is 3.991(1)Å.

In dark red Cu_2Mo_2 derivative [152] the catecholate subunits are cofaxially oriented. The η^1-O_1; η^1;μ-O_2 bridging mode of the catecholate ligands accounts in pair for the "hairpin" structure of the $[Mo_2O_5\{Cu(EtO)_2(\mu\text{-salphen})_2]^{2-}$ anion (Table 4).

In dark $[Cl_3Zn(\mu\text{-Cl})Cu(\mu\text{-enp})]_2$ derivative [153] the four-membered Cu_2O_2 bridging unit is strictly planar owing to the inversion symmetry. Between Cu(II) and Zn(II) there is a single bridging chlorine atom. Each copper(II) atom shows distorted 4 + 1 + 1 square-bipyramidal coordination (CuO_3N_2Cl) and Zn(II) atoms are tetrahedrally coordinated ($ZnCl_4$).

Structure of blue Cu_2Cd_2 derivative [154] is shown in Fig. 8. The metal atoms are bound through asymmetric oxo bridges (Cu - O = 1.927(3) and Cd - O = 2.266(3)Å. Additionally both Cd(II) atoms are bridged by two symmetric oxo bridges (Cd - O = 2.319(3)Å). The six oxygen bridged atoms come from two deprotonated molecules of 1,3,5-tris(trifluoromethyl)-1,3,5-trioxycyclohexane located above and below the plane of each ligand act as bridges between Cd(II) and Cu(II) atoms and the third bridges the two Cd(II) atoms. Each Cd(II) is further coordinated by a bidentate hexafluoroacetylacetonate ligand and each Cu(II) by a bidentate 1,10-phenantroline. The four metal atoms are co-planar, with the intermetal distances Cu - Cd 3.3557(6)Å, Cd - Cd 3.4927(6)Å and Cu - Cu 5.731(1)Å.

Pale blue Cu_2Hg_2 derivative [155] has a pair of copper(II) atoms bridged by two methoxide ions. Each Cu(II) atom has a square pyramidal coordination geometry formed by five oxygen atoms, two from methoxide ions, one from NO_3^-, and one from each of two different $Hg(mp)_2$ units. These last two provide additional bridges between the basal and apical coordination sites of the bridged Cu(II) atoms and, in so doing, form 10- membered macrocyclic rings. The Cu_2O_2 ring is planar with Cu - O - Cu angle of 101,8(8)° and Cu - Cu distance of 2.957(6)Å. Each Hg(II) atom has an almost linear coordination, HgN_2, with N - Hg - N bond angle of 174.6(9)°. The Cu...Hg separation is 3.180(4)Å.

Dark green Cu_2La_2 derivative [58] consists of two binuclear units of [Cu(bdmmpH)(μ-bdmmp)(μ-OH)(F_3ac)La}$_2$ related by an inversion center symmetry. These two units are linked by four trifluoroacetate anions with the mean La - O bond distance of 2.533(9)Å. The Cu(II) atom is in a distorted trigonal bipyramidal environment (CuO_3N_2) and La(III) atom is eight-coordinated (LaO_7N) (Table 4).

Deep blue Cu_2Dy_2 derivative [158] consists of Cu(saltn) and $Dy(hfac)_3$ moieties held together by a double bridge between the metal atoms, the bridging atoms being the oxygen atom of OH group and the oxygen atom of saltn ligand. The OH group is also bonded to the Dy(III) atom related to the inversion center in such a way that the dimeric molecules [Cu(μ_3-OH)(μ-saltn)Dy(hfac)$_3$]$_2$ are formed. The copper(II) atom has a square-planar environment (CuO_2N_2) and dysprosium(III) atom is nine-coordinated (DyO_9).

There is a series of six Cu_2Ln_2 derivatives [157] and each of them contains a central Cu_2O_2 ring bridged to peripheral lanthanoid atoms by 6-methyl-2-pyridone ligands. For lantanoids (Gd, Dy and Yb) the stoichiometry is [(MeOH)(NO_3)$_2$(mhpH)Ln(μ-mhp)$_2$Cu(μ-MeO)]$_2$, and the Ln atoms are eight-coordinate (LnO_8) with a geometry of dodecahedron. For lantanoids (La, Ce and Nd) the complex has a stoichiometry [(MeOH)$_2$(NO_3)$_2$(mhpH)Ln(μ-mhp)$_2$Cu(μ-MeO)$_2$]$_2$, in which the Ln atoms are nine-coordinate (LnO_9), with approximate tricapped trigonal prismatic geometry.

In each case the copper(II) atoms are coordinated by a pairs of oxygen and nitrogen donor atoms in a distorted square-planar environment. The Cu-Cu distance range from 2.956(2) to 2.980(1)Å (Table 4).

Structure of light blue Cu_2Y_2 derivative [107] consists of four-side Cu_2Y_2 unit having Y - Cu distances of 3.317(3) and 3.471(2)Å. The molecule lies over a crystallographic center of inversion. Each of the Y atoms carry two β-diketonate ligands. Metal centers are glued together by 2-methoxyethoxide moieties, and the internal oxygens of these ligands seek additional coordinate to a neighbouring Cu center. There are also two alkoxides which triply bridge a Y and two Cu atoms are further bind to Y through their methoxy functions. While Cu(II) atoms are five-coordinated (CuO_5 - square-pyramidal) the Y(III) atoms are eight-coordinated (YO_8) (Table 4).

Blue green $CuMo_3$ derivative [158] is only example, which contain one copper(II) and another metal atoms. Structure of $CuMo_3O_8(csh)_2(bpy)_2$ is shown in Fig. 9. The structure confirms the presence of a trinuclear unit $[Mo_3O_8]^{2-}$ coordinatively bound to $[Cu(bpy)_2]^{2+}$ cation. The Mo(IV) part of the structure was described in terms of three oxo-bridged centers, alternating one tetraoxomolybdate $[MoO_4]^{2-}$ and two six-coordinated MoO_2 (Schiff base) moieties. The two oxo-groups of the capping $[MoO_4]^{2-}$ unit interact with the Schiff base coordinated Mo atoms. The third oxo group of the tripodal $[MoO_4]^{2-}$ tetrahedron bridges the $[Mo_3O_8]^{2-}$ unit to the $[Cu(bpy)_2]^{2+}$ cation, possessing one site in the equatorial plane of the trigonal bipyramidal Cu(II) coordination environment (CuN_4O).

Inspection of the data in Table 4 reveals that there are copper(II) atoms four- (mostly square-planar), five- (mostly square-pyramidal) and six- coordinated, from which the first two by far prevail.

The M - M distances are summarised in Table 4A, where the variety of metals which can interact with the borderline copper(II) atom can be seen. In the series of heterotetranuclear derivatives, the shortest Cu - Cu distance is 2.956(2)Å [157], and the shortest Cu - M distance is 3. 180(4)Å (for M = Hg(II)) [155].

6. HETEROPENTA- AND HETEROOLIGONUCLEAR COPPER(II) COMPOUNDS

The crystallographic and structural data for these compounds are summarized in Table 5., listed and referenced in the order of increasing nuclearity and there of decreasing number of copper(II) atoms.

There are nine heteropentanuclear derivatives Cu_4M, M = Na [159];or Ba [160, 161]; Cu_3M_2, M = Zn [162]. or Gd [163], CuM_4, M = Tc [164] or Fe [165] and Cu_2Na_2Pd [166].

The structure of a green Cu_4Na consists [159] of two crystallographically independent $Cu_2(\mu\text{-asp})_2(bpy)_2$ molecules, one Na(I) interacting with carboxylate oxygens, a ClO_4^- anion wekly hydrogen bonded to a water molecule, and five water molecules. The binuclear complex molecules result through bridge formation between two copper(II) atoms by the α- and β-carboxylate groups of two aspartate anions. The four copper(II) atoms exhibit severely distorted, but similar square-pyramidal geometries (CuO_3N_2). Two crystallographically independent dimers, "sandwiched" sodium(I) atom, through all carboxylate oxygens which are involved in strong interaction with the Na(I) atom (NaO_8).

In another two green Cu_4Ba derivatives [161] a barium(II) atom is sandwiched between the two $[Cu_2L_2(pyO)_2(F_3CCO_2)]^-$ (L = deae or bdmap) anions. The oxygen atoms on the pyO^-ligands coordinated exclusively to the barium(II) atom. The Ba(II) atom is also coordinated by the additional four oxygen atoms from the deae or bdmap ligands in a nonplanar fashion. Eight oxygen atoms about the barium(II) atom created a distorted square-antiprism geometry. In the two anionic dicopper units two copper(II) atoms are bridged by the two oxygen atoms of the deae or bdmap ligands and in additional only one of the amino groups from each deae or bdmap ligands is coordinated, while the other amino group is away the copper(II) atom. The dicopper unit is capped by a trifluoroacetate ligand. Each Cu(II) center has an approximately square-pyramidal geometry (CuO_3N_2) (Table 5).

The structure of a blue Cu_4Ba complex [161] which has approximate C_2 symmetry, may be viewed as consisting of the binuclear units. They are connected by a Ba(II) atom. Ten oxygen atoms about the Ba(II) atom form a trigonal prism with four capping atoms. Each copper(II) has a pseudo-octahedral geometry (CuO6).

A view of the green Cu_3Zn_2 complex [162] is given in Fig. 10. The pentanuclear $Cu_3(altn)_2(ZnCl_4)_2Cl_2$ molecule is composed of a linear arrangement of three copper(II) atoms, the central one coordinated equatorially by four nitrogen donor atoms from two altn molecules. The axial sites are occupied by chlorine atoms for the central copper(II) atom and one chlorine atom and one $ZnCl_4^{2-}$ anion for both terminal copper(II) atoms. The bridging between the central copper(II) atom and the terminal ones is composed of two triazole-diaza moieties and one chlorine atom. The distance between the central copper(II) atom at the crystallographic center of symmetry and both terminal copper(II) atoms is Cu(1)...Cu(2) 3.620(3)Å.

The structure of blue Cu_3Gd_2 complex [163] consists of centrosymmetrical heteropentanuclear $[Cu_3Gd_2(ceac)_{12}(H_2O)_8]$ molecule. Each terminal Cu(II) atom is coordinated by four carboxy oxygen atoms at the basal plane and by an aqua ligand at the apical position to form a square pyramid. Each terminal Cu(II) is quadruply connected to Gd(III) via four *syn,syn*-acetato-O,O′ bridges with a Cu(II...Gd(III) separation of 3.561(1)Å. A pair of the Gd(III) atoms in two centosymmetricaly related Cu - Gd subunit are each linked to the central Cu(2) atom by a single *syn,anti*-carboxylato-O,O′ group resulting in a pentanuclear molecule. The Cu(2)...Gd(1) distance of 4.662(1)Å is much longer than the Cu(1)...Gd(1) distance (3.561(1)Å).

In $CuTl_4$ derivative [164] the "sunrise" conformation of the four $Tl(18\text{-crown-}6)^+$ cations are facially bound to the central $CuBr_4^{2-}$ ligand. Each Tl(I) atom is seven-coordinated, TlO_6Br. There are two crystallographically independent molecules (Table 5).

Structure of a yellow $CuFe_4$ derivative [165] is shown in Fig. 11. The central copper(II) atom is in a 4+2 surrounding. The four nearest neighbours, at about 2.05Å, are the four nitrogen atoms N(1), N(2), N(1)′and N(2)′from the two diferrocenyl-containing groups which act as bidentate ligands through the ethylenediamine nitrogen atoms. The two apical positions are occupied by oxygen atoms of two nitrate anions (Cu - O = 2.515(6)Å). The ferrocenyl groups are in an eclipsed conformation with twisting angles of 1.6(6) and 4.5(7)° at Fe(1) and Fe(2), respectively.

X-ray structure of $[\{(H_2O)Cu(\mu\text{-wil})_2Na(H_2O)\}_2Pd].9H_2O$ shows [166] that each copper(II) atom is coordinated by D- and L-willardiine cis to each other equatorially and by one water molecule in an apical position (CuO_3N_2). The palladium(II) atom is coordinated in a square-planar geometry, four uracil moieties of the wil molecules also coordinating to copper. Each sodium(I) atom is coordinating to four uracil carbonyl oxygens and one water molecule. The mean Pd-Na distance is 2.976(5)Å. This is only example in the series of heteropentanuclear derivatives which contains a three different metal atoms, Cu_2Na_2Pd.

There are seven heterohexanuclear derivatives: Cu_5Nd [167], Cu_4M_2 (M = Sr [51] or Pr [149, 168]), Cu_2M_4. (M = Fe [169], Co[170] or Ga [171]).

X-ray analysis of a purple Cu_5Nd derivative shows [167] that the cluster contains chiral 15-metallocrown-5 unit, in which the methyl groups of the 2-alaninehydroxamate are all present on only one face of the metallocrown. In addition to the five ring oxygens, the Nd(III) atom is coordinated by one oxygen atom on one face and three oxygens on the other face, each from water molecules. This results in a Nd(III) atom that is nine-coordinate. The Nd(III) atom is significantly displaced from the ring by 0.64Å toward three oxygen atoms of water molecules. The copper(II) atoms are four-coordinated (CuO_2N_2) (Table 5).

Dark blue Cu_4Sr_2 derivative [51] consist of two CuSr units and one Cu_2 unit with a crystallographically imposed inversion center symmetry. In the CuSr unit the metal atoms are bridged together by two bdmap ligands. Most interestingly, this CuSr unit is linked to a Cu_2 unit by a hydroxo group and a bdmap ligand, producing a zig-zag SrCu - Cu_2 - CuSr chain, with Sr...Cu separations of 3.456(4) and 3.457(4)Å. The two hydroxo groups with the Cu - Cu separation of 2.865(5)Å and Cu - O - Cu angle of 93.5(7)°. Each copper(II) atom has a square-pyramidal geometry (Table 5).

Structure of a green complex [149] consists of four Cu(II) atoms and two Pr(III) atoms occupying opposite vertices of the octahedron. The Pr(III) and the Cu(II) are linked together by an oxo ligand (μ_6-O) and eight oxygen atoms from the 2,2′-thiodiethanolate ligands. The Cu...Pr distances range from 3.285(1) to 3.408(1)Å and Cu...Cu distance from 2.7674(13) to 2.980(2)Å. There are three doubly deprotonated tde and two singly deprotonated tdeH ligands. The two Cu(II) atoms have a distorted square-planar geometry (CuO_3S) and remaining two Cu(II) atoms an approximate trigonal bipyramidal geometry (Table 5).

Structure of a yellow complex [168] consists of Cu_4Pr_2 entities in which the metal atoms form a chair-shaped hexagon. The two Pr(III) atoms are located on both sides of a double layer containing the four copper(II) atoms. The copper(II) environments are elongated distorted octahedra (CuO_4N_2), and the praseodymium environments are bicapped square antiprism. (PrO_{10}).

A green $[(thf)Cu(\mu\text{-}O_2CC_5H_4FeCp)_2]_2$ complex [169] has a strict, crystallographically imposed, inversion center. Each copper(II) atom is linked to oxygen atoms from four distinct carboxylate groups and one thf molecule in the form of a square -pyramidal with the copper(II) atom displaced from the basal plane toward the apical thf molecule by 0.182(1)Å. The Cu - Cu distance is 2.605(1)Å. The two (pairs) crystallographically independent ferrocenyl systems have different conformations, two of these have an eclipsed conformation and two have a staggered arrangement.

Structure of dark crimson Cu_2Co_4 derivative [170] consists of the centrosymmetric hexanuclear $[\{Co(dien)(cnde)\}_2Cu(H_2O)_2]^{4+}{}_2$ cations (Fig. 12), ClO_4^- anions, and crystal water molecules. The central core of the complex is a classic lantern, $Cu_2(RCOO)_4(H_2O)$, with Cu - Cu distance of 2.614(2)Å. Each Co(III) atom is six- coordinated (CoN_4O_2) (Table 5).

In a purple Cu_2Ga_4 derivative [171] three mpz ligands and μ_3-O serves as bridges between the metal atoms and contains a planar centrosymmetric Cu_2N_2 ring, with Cu...Cu separation of 3.383(1) and Cu - N - Cu bridge angles of 86.0(1)°. Each copper(II) atom has a distorted square-pyramidal geometry (CuN_4O). Within this same general plane there are two gallium(III) atoms and two pyrazolyl rings, giving a system of four fused five-membered rings in each of the two polycyclic planar systems. The two "planes" making up the centrosymmetric dimer are held together primarily by two stradding "pyrazolyl" moieties which provide a Ga - N - N - Cu link-up between the planes. There are two types of four-coordinate tetrahedral Ga(III) atoms, GaN_2OC and GaC_2ON.

A purple Cu_6Fe derivative [172] is only example, of heteroheptanuclear. The structure (Fig. 13) shows the encapsulation of the $[Fe(CN)_6]^{4-}$ core by six $[Cu(tpma)]^{2+}$ units, forming an almost spherical complex with Cu - Fe distance range from 4.899(4) to 4.929(4)Å. Each Cu(II) atom has a distorted square-planar geometry (CuN_4) and Fe(II) atom an octahedral (FeC_6) (Table 5).

The molecular structure of a blue Cu_4Zr_4 derivative [173] consists of a planar $Cu_4O(O^iPr)_2{}^{4+}$ fragment capped by two $Zr_2O(O^iPr)_8{}^{2-}$ face-sharing bioctahedra units (Fig. 14). The resulting intersecting planes have a dihedral angle of 52.4(3)°. The central oxo of the copper fragment is rigorously square-planar.

The copper coordination environment is essentialy square-planar (CuO_4). The Cu - Cu distance is 2.781(8)Å. The Zr(IV) atoms have distorted octahedra environments (ZrO_6) Table 5).

Dark green Cu_8Y_2cluster [174] is only example of heterodecanuclear for which structure is shown in Fig 15. The molecule has an approximate C_{2h} symmetry. Half of the molecule is related to the other half by an operation of a center of symmetry. There are two distinct groups of dicopper units. One group consists of Cu(1), Cu(2), Cu(1) and Cu(2)′ and the pairs are bridged by two oxygen atoms on the pyo$^-$ ligands. In addition, Cu(1), Cu(2), Cu(1)′ and Cu(2)′ are bonded to one water molecule (Cu(1)-O(14) = 2.33(2)Å), and an oxygen atom on the nitrate ligand (Cu(2) - O(8) = 2.35(3) Å), respectively. As a result, the copper(II) atoms in this group have a square-pyramidal geometry. The separation of copper atoms in the dimeric unit, Cu(1) - Cu(2) is 3.138(7)Å. The other dimeric copper group consists of Cu(3), Cu(4), Cu(3)′ and Cu(4)′. In this group, Cu(3), Cu(4) or Cu(3)′ and Cu(4)′ are bridged by an oxide ligand and a chlorine atom with a Cu(3) - Cu(4) separation of 3.029(7)Å. In this unit, each copper(II atom has a distorted trigonal-bipyramidal geometry (CuO_3NCl). The most importance feature in this molecule is the central core unit of Y_2Cu_4. In this unit, there are two oxide ligands, each of which bridges two Y(III) atoms and two copper(II) atoms. The Y...Y′separation is 3.785(8)Å.

Structure of dark red Cu_2Co_9 derivative [175] is shown in Fig. 16, and it is only example which contains eleven metal atoms. Each metal atom is five-coordinated, CuO_5 and CoC_5, respectively.

Blue $Cu_2W_3Mo_7$ cluster [176] is only example which contains twelve metal atoms and its structure is shown in Fig. 17. As can be seen the P atom is in the center of the anion. The four O atoms of the PO_4 tetrahedron are disordered. The twelve MO_6 octahedra are divided into two classes. Four octahedra link to one O atom of the PO_4 tetrahedron, and the other eight MO_6 octahedra link to two O atoms of the PO_4 tetrahedra, respectively. Every O atom of PO_4 is related to five MO_6 octahedra. The W, Mo and Cu atoms are also disordered in the anion.

The structural units of an orange crystals are $[Cu_2(H_2O)_6WMo_{12}O_{42}]^{4-}$ polyanions, Me_3N^+cations, and water molecules. In the complex anion the central uranium-molybdenum anion coordinates two Cu(II) atoms along a 3 axis, suplying each of them three terminal oxygen atoms in three MoO_2 groups. Three water molecules completed an octahedra environment about each copper(II) atom (Table 5).

There are two blue $Cu_{12}M_6$ (M = Gd or Sm) clusters [178] which are isostructural. Structure of $Cu_{12}Gd_6$ is shown in Fig. 18.

In each structure six Ln(III) atoms are positioned at the vertices of an octahedron with twelve Cu(II) at the midpoints of the edges of the octahedron. The twenty four OH^- groups act as μ_3-bridges each ligating one Ln(III) ant two Cu(II) atoms. The octahedron encapsulates an unusual μ_{12}-ClO_4 anion of which each oxygen atom coordinates to three Cu(II) at the axial positions. Ten of the pyp ligands act in the μ-carboxylato-O,O′ mode, while the other act in the carboxylato-O,O′mode each chelating a Ln(III), resulting in different coordination environments for the Cu(II) atoms (CuO_6) and Ln(III) atoms (LnO_9) (Table 5).

Inspection of the data in Table 5 reveals that there are nine derivatives, which contain five metal atoms, seven derivatives contain six metal atoms, one example of each hepta-, octa-, deca-, undeca- and pentadecanuclear; and two clusters which contain eighteen metal atoms. Copper(II) occuring in four-coordinated environments [164,167,172,173] five- coordinated [159,160,166,51,169,170,171,174,175], six-coordinated [161,162,165,168,176,178], mixed four- and five- coordination [163,149]. There are two examples [159,164] which contain two crystallographically independent molecules differ mostly by degree of distortion and they are another examples of distortion isomers.

Metal-metal distances are summarized in Table 5A. The shortest Cu(II) - M distance is 3.034(2) Å (M = Mo(VI) [177]) and Cu(II) - Cu(II) distance is 2.605(1) Å [169].

7. HETEROPOLYNUCLEAR COPPER(II) COMPOUNDS

Almost one hundred and twenty compounds are found in this category and are listed in Table 6 in the order of increasing coordination number, increasing complexity of the coordination sphere, and increasing atomic number of the principal coordinating ligand about copper(II) atom. The copper(II) coordination number ranges from four to eight, with the "borderline" nature of the copper(II) atom apparent from the structures. In principal the structures are very complex. There are ower thirty derivatives [179-203], in which copper(II) atoms are four-coordinated mostly square-planar with a different degree of distortion.

Structure [Cu(μ-obbz)Mn(H_2O)$_3$].dmf [119] is shown in Fig. 19 as an example of this group. The structure of the compound consists of alternating CuMn chains running along the c axis with Cu(II) and Mn(II) atoms bridged by oxamido (Cu...Mn = 5.483(1)Å) and carboxylato (Cu...Mn = 4.608(1)Å) groups. The conformation around the carboxylato bridge is of the *syn-anti* type. Each unit along the chain is symmetry related to its neighbour by the 2-fold screw axis.

The environments about copper(II) atoms in these compounds are: CuO_4 [179-189], CuN_4 [190, 191], CuS_4 [192], CuO_3N [193] and CuO_2N_2 [194-203].

There are three examples [186,189,200] which contain non-equivalent copper(II) atoms, a four- and five-coordinated, and one example [191] in which copper(II) atoms are four- and six- coordinated (Table 6).

In the series of heterometallic copper(II) compounds in which copper(II) atoms are four-coordinated, another metal atoms is non- transition, transition or lanthanide. Stereochemistries about non-transition metals is in quite wide range trigonal, tetrahedral, trigonal-bipyramidal, a pseudo-octahedral, pentagonal-bipyramidal, dodecahedral and a capped square-antiprismatic. Stereochemistry about transition metals is; a square-planar, a pseudo-octahedral, and pentagonal-bipyramidal. Stereochemistry about Tm(III) atom [203] (this is only example of lanthanoide) is a capped trigonal prismatic (TmO_7).

There are almost thirty derivatives [75,197,204-226] in which copper(II) atoms are five- coordinated. Structures are very complex, and X-ray structure of a green $Cu_2(MoO_2)_2$(saladhp)$_2$(MeO)$_2$ [212] is shows in Fig. 20 as an example. The central cubane-like core of the compound can be described as a strongly distorted cube in which four corners are occupied by two copper(II) and two molybdenum(VI) atoms. The cube is completed by four oxygen atoms from the two Schiff base molecules and two methoxy ligands. The copper atoms are bridged by the alkoxy oxygen atoms of the saladhp ligand with a Cu - Cu distance of 2.951(5)Å, and the molybdenum atoms by the methoxy oxygen atoms with a Mo...Mo distance of 3.528(4)Å. The Mo and Cu atoms are connected by one methoxy and one alkoxy oxygen atom with an average

Cu...Mo distance of 3.429Å. The O(3) atom of the Mo = O moiety of each core has a contact to a neighbouring cubane through its Cu(2) atom, affording an infinite chain arrangement. The coordination polyhedron around each Mo(VI) atom is a distorted octahedron (MoO_6). The Cu(1) atom is in a distorted square-pyramidal environment (CuO_4N) and the Cu(2) atom in a distorted octahedral (CuO_5N) (Table 6).

In this group, copper(II) atoms (mostly in a square-pyramidal environment) are surrounded by a donor atoms: CuO_5 [204-206], CuN_5 [207-210], CuO_4N [211,212], CuN_4O [197,210,213], CuO_3N_2 [197, 215-222], CuN_3O_2 [223,224], CuN_3OCl [225], CuO_2N_2S [226] and CuO_2N_2Br [75]. In this shies of a square-pyramidal environments around Cu(II), the displacement of the Cu(II) atom from the square plane donor atoms towards the apical ligand range from 0.032 Å to 0.32 Å.

Besides copper(II) atoms in this series another metal is non-transition which are three-, four-, five- and six- coordinated, transition which are four-, and six-coordinated; and lanthanoides which are eight-, nine- and ten-coordinated (Table 6).

The $Cu(\mu\text{-}Sb)Mn(CO)_3$ [224] contains two crystallographically independent molecules differ mostly by degree of distortion and it is another example of distortion isomers.

There are over fifty derivatives [30, 227-271] which contain a six- coordinated copper(II) atom. Structure of a blue $Sr_2Cu(ox)_3(H_2O)_7$ [235] is shown in Fig. 21, as an example. The structure can be described as a complex three-dimensional network of Cu(II) and Sr(II) atoms coordinated by oxalate and water molecules. The coordination polyhedra around the Cu(II) atoms a distorted octahedra (CuO_6) and the Sr(II) atoms are nona-coordinated (SrO_9) (Table 6).

The most common donor atoms are O and N, which created an octahedral environment about copper(II) with a tetragonal-bipyramidal being the most common geometry with the chromophores: CuO_6 [227-241], CuN_6 [242-245], CuO_5N [246, 247], CuN_5O [248], CuO_4N_2 [249-257], CuO_4Cl_2 [258], CuN_4F_2 [259], CuN_4O_2 [260-264], CuN_4Cl_2 [30, 265], CuN_4Br_2 [265], CuN_4S_2 [266], $CuCl_4O_2$ [267], CuO_3N_3 [268], CuO_4NI [269], CuO_3N_2F [259], $CuO_2N_2Cl_2$ [270] and $CuO_2N_2Br_2$ [271]..

Inspection of the data in Table 6 reveals that the multidentate ligands include both homo- and hetero-donor atoms, with O-, N-, and O plus N- donors being by far the most common. The effect of both electronic and steric factors are observed in the values of the L - Cu - L bond angles of the metallocyclic rings. There is interdependence between the metallocyclic ring and the coordination number of copper(II) atom, the angle is closses with an increases of the coordination number (c.n.). In five- membered rings the mean L - Cu - L intra-ligand angles are: 80.2° (c.n. 6) < 83.7° (c.n. 5) < 86.8° (c.n. 4) (O-donor); 83.5° (c.n. 6) < 84.0° (c.n. 5) < 86.0° (c.n. 4) (N-donor); 78.5° (c.n. 6) < 83.5° (c.n. 5) < 85.5° (c.n. 4) (O- plus N-donor); In six- membered rings the mean values: 98.0° (c.n.6) < 96.0° (c.n. 5) (N-donor), and 91.0° (c.n. 6) < 93.0° (c.n. 5) < 93.5° (c.n. 4) (O- pus N-donor).

The metal-metal separations found in heteropolynuclear copper(II) compounds are summarised in Table 6A. The shortest Cu(II) - M distances found in this series are; with non-transition Cu - Na = 3.138(2) [251], with transition Cu - Fe = 3.033(2) [200], and with lanthanoide Cu - Ce = 3.288(2) [220]. The shortest M - M distances are: Li(I) - Li(I) = 2.529(8) [181] and Cu - Cu = 2.633(1) [186].

There are two derivatives [272-274] in which copper(II) atoms are eight coordinated. The crystal structure of blue $[Cu(\mu\text{-bet})_4Li](ClO_4)_3$ [272] featuring an infinite cationic column in which each pair of adjacent Cu(II) and Li(I) atoms are doubly bridged by a pair of bet ligands, with disordered perchlorate anions filling the intercolumnar space in the crystal packing. The Cu(II) atom, occupying Wyckoff position 1(a) of site symmetry *42m* (D_{2d}), is coordinated by four carboxylate groups of the bet ligands with four short Cu - O bonds (1.970(4)Å) in an idealized square-planar fashion. The coordination sphere of the metal atom is completed by the remaining four carboxylate oxygen atoms at much longer distances (2.706(5)Å. The betaine ligands, as located at Wyckoff positions 4(n) of symmetry *m*, coordinate the Cu(II) atom in an alternatively up/down fashion, the coordination geometry about the Cu(II) atom is a distorted dodecahedron. Two pairs of the long-bonded oxygen atoms of the trans-related carboxylate groups from adjacent $Cu(bet)_4^{2+}$ cores chelate to a Li(I) ion located at Wyckoff position 1(c) of symmetry *42m*, resulting in tetrahedral coordination about the Li(I) atom (LiO_4). The Cu - Li distance is 3.116(1)Å.

Structure of $[Cu(\mu\text{-ac})_4Ca(H_2O)_2]4H_2O$ was studied by two groups [273, 274]. Each acetate anion is bidentate and acts as a bridging between two metal atoms in such a manner as to produce polymeric chains of alternate metal atoms parallel to the c-axis of the crystal. The Cu(II) atom occupies the *4* (S_4) site and is coordinated to a puckered square of four oxygen (O(2)) atoms at 1.97Å. The O(1) atoms which are coordinated to calcium are bridging between Ca(II) and Cu(II) and are 2.79Å from Cu. Two of these O(1) atoms lie above the puckered square and two below. The O(2) atoms describe a tetrahedrally distorted square plane while the O(1) atoms describe an elongated tetrahedron about the metal atom. The symmetry of the resulting coordination about the Cu(II) atom is approximately D_{2d} and is a distorted dodecahedron.

The Ca(II) atom occupies a site of *2/m* (C_{2h}) symmetry and is six coordinate with four O(1) atoms from the bridging acetates and two water molecules. The Cu...Ca distance is 4.060(1) Å (Table 6).

8. CONCLUSIONS

The data presented in this review cover some three hundred and sixty heteronuclear copper(II) coordination compounds for which the structural data are available. The number of examples of different nuclearity increases in the order: heteropoly- < -bi- < -tri- < -tetra- < -penta- < -hexa- < -deca- ~ -octadeca < -hepta-, -octa-, -undeca- and -pentadecanuclear.

There are mono- and multidentate ligands include both homo- and hetero-donor atoms, with O-, N-, and O plus N- donors being by far the most common.

Two derivatives [12,36] are found in two isomeric forms differing only by degree of distortion. In several derivatives [24,57a,65,93,137,159,164,224] two crystallographically independent molecules are present in the same crystal. The differences occur in both the bond lengths and the bond angles. Such examples are typical of the general class of distortion isomerism [97].

Copper(II) atoms are most often found in distorted square-planar square-pyramidal and tetragonal-bipyramidal environments. There are two examples [272-274] in which copper(II) atoms are in a distorted dodecahedral environments. The Cu - M and M - M distances are summarised in each chapter (Table 6A), where the variety of metals which can interact with the borderline copper (II) atom can be seen.

A similar review of copper(II) heterometallic compounds, has surveyed some two hundred and forty compounds [9].

The present review, together with the others [1-8] analysed and classified over six thousand copper derivatives, shows the richness of the chemistry of copper and the variety of structural arrangements and nuclearity. During the collection and organisation of the data it has became clear that despite the increasing availability of data-retrieval systems, finding relevant information is not always a straightforward task. One of the problems seems to be associated with the choice of key-words for indexes, and some structural material effectively vanishes from a structural search. Some of the data is only available as supplementary material, which may lead to the overlooking of certain relevant features for comparative purposes. It is hoped that reviews such as this might help to both alleviate and perhaps highlight some of these aspects, and at the some time serve a useful purpose by pointing out features and areas of particular interest.

9. ACKNOWLEDGEMENTS

The authors wish to thank those who gave permission for reproduction of original figures, and the Ministry of Education of Slovakia for financial support.

10. REFERENCES

1. C.E. Holloway and M. Melník, *Rev. Inorg. Chem.,* **15,** 147 (1995).
2. M. Melník, M. Kabešová, M. Dunaj-Jurčo and C.E. Holloway, *J. Coord. Chem.,* **41,** 35 (1997).
3. M. Melník, M. Kabešová, Ľ. Macášková and C.E. Holloway, *J. Coord. Chem.,* **45,** 35 (1997).
4. M. Melník, M. Kabešová, M. Koman, Ľ. Macášková, J. Garaj, C.E. Holloway, and J.Valent *J. Coord. Chem.,* **45,** 147 (1997).
5. M. Melník, M. Kabešová, M. Koman, Ľ. Macášková and C.E. Holloway, *J. Coord.Chem.,* in press.
6. M. Melník, M. Kabešová, M. Koman and C.E. Holloway, *J. Coord. Chem.,* in press.
7. M.Melník and M. Kabešová, *J. Coord. Chem.,* in press
8. M. Dunaj-Jurčo, G. Ondrejovič, M. Melník and J. Garaj, *Coord. Chem. Rev.,* **83,** 1 (1988).
9. C.E. Holloway and M. Melník, *Rev. Inorg. Chem.,* **13,** 233 (1993).
10. C.P. Love, Ch.C. Torardi and C.J. Page, *Inorg Chem,* **31,** 1784 (1992).
11. B. Morosin, P. Fallon and J.S. Valentine, *Acta Cryst., Sect. B,* **31,** 2220 (1975).
12. H. Oshio, K. Toriumi and S. Bandow, *Chem. Let.,* 1713 (1998); H. Oshio, K. Toriumi, S. Bandow, K. Miyagawa and S. Kurita, *J. Chem. Soc., Dalton Trans.,* 1013 (1990).

13a. A.M. Duda, A. Karaczyn, H. Kozlowski, I.O. Fritsky, T. Glowiak, E.V. Prisyazhnaya, T.Yu. Sliva and J. Swiatek-Kozlovska, *J. Chem Soc., Dalton Trans.*, 3853 (1997) .

13b. W.R. Scheidt, B. Cheng, K.J. Haller, A. Mislankar, A.D. Rae, W. Reddy, H. Song, R.D. Orosz, Ch.A. Reed, F. Cukiernik and J.C. Marchan, *J. Am. Chem. Soc.*, **115**, 1181 (1993).

14. K.P. Larsen, R. Gronbek Hazell, H. Toflund, P.R. Andersen, P. Bisgård, K. Edlund, M. Eliasen, C. Herskind, T. Laursen and P. Moller Pedersen, *Acta Chem. Scand., Ser A*, **29**, 499 (1975).

15. N.R. Strel'tsova, V.K. Bel'skii, B.M. Bulychev and O.K. Kireeva, *Zh. Neorg. Khim.*, **37**, 1822 (1992); Engl. ed. p. 934

16. T.B. Rubstova, O.K. Kireeva, B.M. Bulychev, N.P. Strel'tsova, V.K. Bel'skii and B.P. Tarasov, *Polyhedron*, **11**, 1929 (1992).

17a. B.M. Bulytchev, O.K. Kireeva V.K. Bel'skii and N.R. Strel'tsova, *Polyhedron*, **11**, 1809 (1992).

17b. J.F. Biernat, A. Cygan, E. Luboch, Y.A. Simonov, T.I. Malinovskij, V.K. Bel'skii and N.F. Bolotina, *J. Inclusion Phenom. Mol. Rec. Chem.*, **15**, 369 (1993).

18. M. Fettouhi, L. Ouhab, M. Hagiwara, E. Codjovi, O. Kahn, H. Constant-Machado and F. Varret, *Inorg. Chem.*, **34**, 4152(1995).

19. A.Bino and M. Cohen, *Inorg. Chim. Acta*, **141**, 5 (1988).

20. P.A. Chetculi, A. Liégard, G. Ribs, G. Rist and A. Schweiger, *Helv. Chim. Acta*, **74**, 1591 (1991).

21. I. Bernal, J.D. Korp, E.O. Schlemper and M.S. Hussain, *Polyhedron*, **1**, 365 (1982).

22. D.J. Hodgson, P.K. Hale and W.E. Hatfield, *Inorg. Chem.*, **10**, 1061 (1971)

23. S.K. Hofmann, D.J. Hodgson and W.E. Hatfield, *Inorg. Chem.*, **24**, 1194 (1985).

24. S. Kotila and J. Valtonen, *Acta Chem. Scand.*, **48**, 200 (1994)

25. S. Mahapatra, J.A. Halfen, E.C. Wilkinson, G. Pan, X. Wang, V.G. Yong Jr., Ch. J. Cramer, L.Que Jr., and W.B. Tolman, *J. Am. Chem. Soc.*, **118**, 11555 (1996).

26. J. Fischer, R. Elchinger and R. Weiss, *Acta Cryst., Sect.B*, **29**, 1967 (1973).

27. S. Gorter and G.C. Verschoor, *Acta Cryst., Sect. B*, **32**, 1704 (1976).

28. H. Oshio, *Inorg. Chem.*, **32**, 4123 (1993).

29a. G.De Munno, C. Bazzicalupi, J. Fous, F. Lloret and M. Julve, *J. Chem. Soc., Dalton Trans.*, 1879 (1994).

29b. P.G. Jones, *Acta Cryst., Sect. C*, **48**, 1314 (1992).

30. J. Pickardt and I. Hofmeister, *Z. Naturforsch.*, **50b**, 828 (1995).

31. H. Werner, J. Roll, K. Linse and M.L. Ziegler, *Angew. Chem. Int. Ed. Engl.*, **22**, 982 (1983); H. Werner, J. Roll, R. Zolk, P. Thomeczek, K. Linse and M.L. Ziegler, *Chem. Ber.*, **120**, 1553 (1987).

32. D. Neugebauer and B. Lippert, *J. Am Chem. Soc.*, **104**, 6596 (1982).

33a. C. Sirio, O. Pancelet, L.C. Hubert-Pfalzgraf, J.C. Daran and J. Vaissermann, *Polyhedron*, **11**, 177 (1992).

33b. Y. Journaux, O. Kahn, I. Morgenstern-Badarau, J. Gely, J. Jound, A Bencini and D. Gatteshi , *J. Am. Chem. Soc.*, **107**, 6305 (1985).

33c. Fung Lam, Ru-Ji Wang, T.C.W. Mak and Kin Shing Chan, *J. Chem. Soc., Chem. Commun.*, 2439 (1994).

34. R.L. Lintvedt, L. Stecher Kramer, G. Ranger, P.V. Corfield and M.D. Glick, *Inorg. Chem.*, **22**, 3580 (1983).
35. R Gracioni, M. Vidali, G. Rizzardi, U. Casellato and P.A. Vigato, *Inorg. Chim. Acta,* **36**, 145 (1979).
36. O.Kahn , J.Galy, Y. Journaux, J. Jand and I. Morgenstern-Baderau, *J. Am. Chem. Soc.,* **104**, 2165 (1982).
37. F. Teixidor, T. Flor, J. Casabó, C. Miravitlles, J. Rius and E. Molins, *Inorg. Chim. Acta,* **122**, 95 (1986).
38. I. Morgerstern- Baderau, M. Rerat, O. Kahn, J. Jand and J. Galy, *Inorg. Chem.*, **21**, 3050 (1982).
39. H. Adams, N.A. Bailey, D.E. Fenton, M.S.L. Gonzales and C.A. Philips, *J. Chem. Soc., Dalton Trans.*, 371 (1983).
40. C. Fraser, R. Ostrander, A.L. Reingold, C. White and B. Bosnich, *Inorg Chem.*, **33**, 324 (1994).
41. J.P. Beale, J.A. Cuningham and D.J. Philips, *Inorg. Chim. Acta,* **33**, 113 (1979).
42. G.A. Brewer and E. Sinn, *Inorg. Chem.*, **26**, 1529 (1987).
43. M.Mikuriya, H. Otawa, S.Kida and I. Neda, *Bull. Chem. Soc. Jpn.*, **51**, 2920 (1978).
44. J. Galy, J. Jand, O. Kahn and P. Tola, *Inorg. Chem.*, **19**, 2945 (1980).
45. D.E. Fenton, *J. Chem. Soc., Chem. Commun.*, 39 (1979).
46. T. Aono, H. Wada, Y. Aratake, N. Matsumoto, H. Õkawa and Y. Matsuda, *J. Chem. Soc., Dalton Trans.*, 25 (1996).
47. N.B. O'Bryan, T.O. Maier, I.C. Paul and R.S. Drago, *J. Am. Chem. Soc.*, **95**, 6640 (1973).
48a. H. Õkawa, J. Nishio, M. Ohba, M. Tadakoro, N. Matsumoto, M. Koikawa, S. Kida and D.E. Fenton, *Inorg. Chem.*, **32**, 2949 (1993).
48b. R. Ruiz, F. Lloret, M. Julve, J. Faus, M. Carmen Muòoz and X. Solans, *Inorg. Chim. Acta,* **213**, 261 (1993).
49. J. Jaud, Y. Journaux, J. Galy and O. Kahn, *New. J. Chem.*, **4**, 629 (1980); Y. Journaux, O. Kahn J. Zarembovitch, J. Galy and J. Jaud, *J. Am. Chem/ Soc.*, **105**, 7585 (1983).
50. S. Ohtsuka, M. Kodera, K. Motoda, M. Ohba and H. Õkawa, *J. Chem. Soc., Dalton Trans.*, 2599 (1995).
51. S.R. Breeze and S. Wang, *Inorg Chem.*, **33**, 5113 (1994).
52. M. Tadokozo, H. Sakiyama, N. Matsumoto, M. Kodera, H. Õkawa and S. Kida, *J. Chem. Soc., Dalton Trans.*, 313 (1992).
53. Suning Wang, Zhen Pang, K.D.L. Smith, You-sheng Hue, Ch. Deslippe and M.J. Wagner, *Inorg. Chem.*, **34**, 908 (1995); Suning Wang, Zhen Pang and K.D.L. Smith, *ibid*, **32**, 4992 (1993).
54. T.R. Holman, K.A. Andersen, O.P. Anderson, M.P. Hendrich, C. Juarez-Garcia, E. Münck and L. Oue,Jr., *Angew. Chem. Int. Ed. Engl.*, **29**, 921 (1990).
55. A. Tiripicchio, A.M. Manotti Lanfredi, M. Ghedini and F. Neve, *J. Chem. Soc. , Chem. Commun.*, 97 (1983).
56. J.P. Costes. F. Dahan, A. Dupius and J.P. Laurent, *Inorg Chem.*, **35**, 2400 (1996).
57a. E. Colacio, J.M. Dominguez-Vera, A. Escuer, R. Kivekäs and A. Romerosa, *Inorg. Chem.*, **33**, 3914 (1994); E. Colacio, J.M. Dominguez-Vera, A.

Romerosa, R. Kivekäs, M. Klinga and A. Escuer, *Inorg. Chim. Acta,* **234**, 61 (1995).

57b. Feng Gao, Rui-Yao Wang, Tion-Zhu Jin, Guang-Xian Xu, Zhang-Yuan Zhou and Xiang-Ge Zhou, *Polyhedron,* **16**, 1357 (1997).

57c. R. Graziani, M. Vidali, U. Cosellato and P.A. Vigato, *Transition Met. Chem.*, **3**, 239 (1978).

58. Liqin Chen, S.R. Breeze, R.J. Rousseau, Sining Wang and K.L. Thompson, *Inorg. Chem.*, **34**, 454 (1995).

59. F. Birkelbach, M. Winter, U. Flörke, H.J. Haupt, H. Butzlaff, M. Lengen, E. Bill, A.X. Trautwein, K. Wieghart and P. Chaudhuri, *Inorg. Chem.*, **33**, 3990 (1994).

60. M.J. Scot, Hua H. Zhang, S.C. Lee, B. Hedman, K.O. Hodgson and R.H. Holm, *J. Am. Chem. Soc.*, **117**, 568 (1995).

61. I.B. Liss and E.O. Schlemper, *Inorg. Chem.*, **14**, 3035 (1975).

62. R. Cortés, L. Lezama, J.I.R. Larramendi, M. Insausti, J.V. Folgado, G. Madariaga and T. Rojo, *J. Chem. Soc., Dalton Trans.*, 2573 (1994).

63. A. Nanthakumar, S. Fox, N.N. Murthy. K.D. Karlin, N. Ravi, B.H.Huynh, R.D. Orosz, E.P.Day K.S. Hagen and N.J. Blackburn, *J. Am. Chem. Soc.*, **115**, 8513 (1993); K.D. Karlin, A. Nanthakumar, S. Fox, N.N. Murthy, N. Ravi, B.H. Huynh, R.D. Orosz and E.P.Day, *ibid,* **116**, 4753 (1994).

64. J.P.Costes, J. F. Serra, F. Dahan and J.P. Laurent, *Inorg, Chem.*, **25**, 2790 (1986).

65. F.C.J.M. van Vegel, S. Harkema, M. Bos, W. Verboom, G.K. Woolthuis and D.N. Reinhoudt, *J. Org. Chem.*, **54**, 2351 (1989).

66. V.M. Agra, T.F. Sysoeva, V.K. Trunov, A.Yu. Fridman and N.N. Barkhanova, *Zh. Strukt. Khim.,* **22,** 114 (1981) Engl. Ed., p. 86

67. A. Fuertes, C. Miravitlles, E. Molins, E. Escrivá and D. Beltrán, *Acta Cryst., Sect. C.,* **42**, 421 (1986).

68. A. Fuertes, C. Miravitlles, E. Escrivá, E. Coronado and D. Beltrán, *J. Chem. Soc., Dalton Trans.*, 1847 (1987).

69a. B.R. Serr, Ch.E.L. Headford, O.P. Anderson, C.M. Eliot, K. Spartalion, V.E. Fainzilberg, W.E. Hatfield, B.R. Rohrs, S.S. Eaton and G.R. Eaton, *Inorg. Chem.*, **31**, 5450 (1992).

69b. S.C. Lee and R.H. Holm, *Inorg. Chem.*, **32**, 4745 (1993).

70. D.A. Wroblecki and T.B. Rauchfuss, *J. Am. Chem. Soc.*, **104**, 2314 (1982); D.A. Wroblecki, C.S. Day, B.A. Goodman and T.B. Rauchfuss, *ibid,* **106**, 5464 (1984).

71. B.R. Serr, Ch.E.L. Headford, C.M. Eliot and O.P. Anderson, *J. Chem. Soc., Chem. Commun.*, 92 (1988).

72a. Zhuang Jin Zhong, N. Matsumoto, H. Õkava and S. Kida, *J. Chem. Soc. Dalton Trans.*, 2095 (1989).

72b. R. Guilard, S. Brandes, A. Tabard, N. Bouhmaida, C. Lecomte, Ph. Richard and J.M. Latour, *J. Am. Chem. Soc.*, **116**, 10202 (1994).

73. M.J. Scot, Ch.A. Goddard and R.H. Holm, *Inorg. Chem.*, **35**, 2558 (1996).

74a. B. Chiari, A. Cinti, O. Piovesana and P.F.Zanazi, *Inorg. Chem.*, **34**, 2652 (1995).

74b. M.J. Scott, S.C. Lee and R.H. Holm, *Inorg. Chem.*, **33**, 4651 (1994).

75. K. Nakatani, J. Sletten, S. Habut-Desporte, S. Jeannin, Y. Jeannin and O. Kahn, *Inorg. Chem.*, **30**, 164 (1991).
76a. K. Nakatani, J.Y. Carriat, Y. Journaux, O. Kahn, F. Lloret, J.P. Renard, Yu Pei, J. Sletten and M. Verdaguer, *J. Am. Chem. Soc.*, **111**, 5739 (1989).
76b. Dai-Zheng Liao, Li-Cun Li, Zeng-Hui Jiang, Shi-Ping Yan, Peng Cheng, Geng-Lin Wang, Xin-Kan Yao and Hong-Gen Wang, *Transition Met. Chem.*, **17**, 356 (1992).
77. A. Escuer, R. Vincente, J. Ribas, R. Costa and X. Solans, *Inorg. Chem.*, **31**, 2627 (1992).
78. Zhi-Yang Zhang, Dai-Zheng Liao, Zeng-Hui Jiang, Song-Qi Hao Xin-Kan Yao, Hong-Gen Wang and Geng-Lin Wang, *Inorg. Chim. Acta,* **173**, 201 (1990).
79. Yu Pei, O. Kahn, K. Nakatani, E. Codjovi, C. Mathonière and J. Sletten, *J. Am. Chem. Soc.*, **113**, 6558 (1991).
80. C. Mathonière, O. Kahn, J.C. Daran, H. Hilbig and F.H. Köhler, *Inorg. Chem.*, **32**, 4057 (1993).
81. M. Ohba, H. Tamaki, N. Matsumoto and H. Õkawa, *Inorg. Chem.*, **32**, 5385 (1993)
82. Q. Lu, Q.H. Luo, A.B. Dai, Z.Y.Zhou and G.Z. Hu, *J. Chem. Soc., Chem. Commun.*, 1429 (1990).
83. W.M. Davis, J.C. Devan and S.J.Lippard, *Inorg. Chem.*, **20**, 2928 (1981).
84. H. Adams, N.A. Bailey, D.E. Fenton and Qing-Yu He, *J. Chem. Soc., Dalton Trans.*, 1533 (1997).
85. J.L. Pierre, P. Chautemps, S.M. Refail, C.G. Béguin, A.El. Marzouki, G. Serratrice, E. Saint-Aman and P. Rey, *J. Am. Chem. Soc.*, **117**, 1965 (1995); J.L. Pierre, P. Chautemps, S.M. Refail, C.G. Béguin, A.El. Marzouki, G. Serratrice, P. Rey and J. Laugier, *J. Chem. Soc., Chem. Commun.*, 1117 (1994).
86. Zong-Wan Mao, Dong Chen, Wen-Xia Tang, Kai-Bei Yu and Li Liu, *Chin. J. Chem.*, **10**, 45 (1992); *Polyhedron,* **11**, 191 (1992).
87. Zong-Wang Mao, Min-Qin Chen, Xiang-Shi Tan, Jie Liu and Wen-Xia Tang, *Inorg.Chem.*, **34**, 2889 (1995).
88. N. Matsumoto, H. Õkawa, S. Kida, T. Ogawa and A. Ohyoshi, *Bull. Chem. Soc. Jpn.*, **62**, 3812 (1989).
89. X. Bing, Ch. Dang, T. Wenxia, Y. Kaibei and Z. Zhongyuang, *Acta Cryst.,Sect.C,* **47**, 1805 (1991).
90. Ping-Rong Wei, Lu Jia, Chang-Rang Liu, Yao-Ting Fan, and Song Gao, *Chin. J. Chem.*, **12**, 503 (1994).
91. B. Crociani, G. Bandoli and D.A. Clemente, *J. Organomet. Chem.*, **190**, C97 (1980).
92. K.R.J. Thomas, V. Chandrasekhar, C.D. Bryan and A.W. Cordes, *J. Coord. Chem.*, **35**, 337 (1995).
93. K.L.Woo, M.R. Maurya, R.A. Jacobson, S. Yang and S.L. Ringrose, *Inorg. Chem.*, **31**, 913 (1995).
94. O.P. Gladkikh, N.F. Curtis and S.L. Heath, *Acta Cryst., Sect. C,* **53**, 197 (1997).
95. H. Kurihara, S. Nisnikiori and T. Iwamoto, *Acta Cryst., Sect. C,* **53**, 1409 (1997).
96. R. Cortés, M.K. Urtiaga, L. Lezama, M.I. Arriortua and T, Rojo, *Inorg. Chem.*, **33**, 829 (1994).
97. M. Melník, *Coord. Chem. Rev.*, **47**, 239 (1982).

98. G.A. Brewer and E. Sinn, *Inorg. Chim. Acta,* **134**, 13 (1987).
99. A, Bencini, C. Benelli and D. Gatteschi, *Inorg. Chem.*, **24**, 695 (1985).
100. K. Brychey, K.J. Jens, M. Tilset and U. Behrens, *Chem. Ber.*, **127**, 991 (1994).
101. A. Bencini, C. Benelli, A. Caneschi, R.L. Carlin, A.Dei and D. Gateschi, *J. Am. Chem. Soc.*, **127**, 991 (1994).
102. A. Bencini, C. Benelli, A. Caneschi, A.Dei and D. Gateschi, *Inorg. Chem.*,**25**, 572 (1986).
103. M.Tadokoro, H. Õkawa, N. Matsumoto, M. Koikawa and S.Kida, *J. Chem. Soc., Dalton Trans.*, 1657 (1991).
104. F.C.J.M. van Veggel, M. Bos, S. Harkema, H. van der Bovenkamp, W. Verboom, J. Reedijk and D.N. Reinhoudt, *Angew. Chem., Int. Ed. Engl.*, **28**, 746 (1989); F.C.J.M. van Veggel, M. Bos, S. Harkema, H. van der Bovenkamp, W. Verboom, J. Reedijk and D.N. Reinhoudt, *J. Org. Chem.*, **56**, 225 (1991).
105. Ch. Fukuhara, K. Tsuneyoshi, N. Matsumoto, S. Kida and M. Mori, *J. Chem. Soc., Dalton Trans.*, 3473 (1990).
106. W. Bidell, H.W. Bosch, D. Veghini, H.V. Hund and H. Berke, *Helv. Chim. Acta,* **76**, 596 (1993).
107. W. Bidell, J. Döring, H.W. Bosch, H.U. Hund, E. Plappert and H. Berke, *Inorg. Chem.*, **32**, 502 (1993).
108. Suning Wang, Zhen Pang, K.D.L. Smith and M.J. Wagner, *J. Chem. Soc., Dalton Trans.*, 955 (1994); *J. Chem. Soc., Chem. Commun.*, 1594 (1992).
109. H. Oshio, T. Kikuchi and T. Ito, *Inorg. Chem.*, **35**, 4938 (1996).
110. Zhong-Ling Lu, Chun-Ling Duan, Yu-Peng Tian, Zhen-Wei Wu, Jinf-Jing You, Zhong-Yuan Zhou and T.C.W. Mak, *Polyhedron,* **16**, 909 (1997).
111. De-Gang Fu, Bing Xu, Guo-Xiang Wang, Wen-Xia Tang, Kai-Bei YU and Zhong-Yuan Zhou, *Sc. China, Ser. B,* **37**, 906 (1994).
112. K.K.Nanda, S. Mohanta, U. Flörke, S.K. Dutta and K. Nag, *J. Chem. Soc., Dalton Trans.*, 3831 (1995).
113. A.C. Fabretti, A. Giusti, V.G. Albano, C. Castelani, D. Gateschi and R. Sessoli, *J. Chem. Soc., Dalton Trans.*, 2121 (1991).
114. C. Benelli, A.C. Fabretti and A. Giusti, *J. Chem. Soc., Dalton Trans.*, 409 (1993).
115. K.V.P.M. Shafi, M. Nethaji and D.N. Sathyanarayana, *J. Chem. Soc., Dalton Trans.*, 3179 (1996).
116. J. Glecup, P.A. Goodson, D.J. Hodgson, M. H. Lynn and K. Michelsen, *Inorg. Chem.*, **31**, 4785 (1992).
117. P. Galen Lenhert, C.M. Lukehart anf L.T. Warfield, *Inorg. Chem.*, **19**, 311 (1980).
118. D.J. Patmore, D.F Rendle, A. Stou and J. Trotter, *J. Chem. Soc., Dalton Trans.*, 718 (1975).
119. A. Schreiber, O. Krizanovic, E.C. Fusch, B. Lippert, F. Lianza, A. Abinati, S. Hill, D.M.L. Goodgame, H. Stratemeier and M.A. Hitchman, *Inorg. Chem.*, **33**, 6101 (1994).
120. B. Lippert, U. Thewalt, H. Schölhorn, D.M.L. Goodgame and R.W. Rollins, *Inorg. Chem.*, **23**, 2807 (1984).
121. I. Mutikainen, O. Orama, A. Pajunen and B. Lippert, *Inorg. Chim. Acta,* **137**, 189 (1987).
122. A. Erxleben, A. Albinati and B. Lippert, *J. Chem. Soc., Dalton Trans.*, 1823 (1996).

123. G. Frommer, F. Lianza, A. Albinati and B. Lippert, *Inorg. Chem.*, **31**, 2434 (1992).
124. G.A. van Albada, R.A.G. de Graaf, R. Hage, J. Reedijk, E. Buchholz and W. Kläui, *Polyhedron,* **10**, 1091 (1991).
125. V. Manivannan, S. Dutta, P. Basu and A. Chakravorty, *Inorg. Chem.*, **32**, 4807 (1993).
126. F. Birkelbach, T. Weihermüller, M. Lengen, M. Gerdan, A.X. Trautwein, K. Wieghardt and P. Chaudhuri, *J. Chem. Soc., Dalton Trans.*, 4529 (1997).
127. P. Chaudhuri, M. Winter, P. Fleischhauer, W. Haase, U. Flörke and H.J. Haupt, *J.Chem. Soc., Chem. Commun.*, 1728 (1990).
128. A. Holzmann, P. Berges, W. Hinricks and G. Klar, *J. Chem. Research(S),* 42 (1987).
129. M. Veith, S. Mathur and V. Huch, *Inorg. Chem.*, **36**, 2391 (1997).
130. B.A. Vaartska, J.A. Samuels, E.H. Barash, J.D. Martin, W.E. Streib, Ch. Gasser and K.G. Caulton, *J. Organomet. Chem.*, **449**, 191 (1993).
131. F. Labrize, L.G. Hubertpfalzgraf, J. Vaissermann and C.B. Knobler, *Polyhedron,* **15**, 577 (1996).
132. V.M. Agre, T.F. Sysoeva, V.K. Trunov, V.A. Efremov, A.Ya. Fridman and N.N Barkhanova, *Zh. Strukt. Khim.*, **2**, 110 (1980); Engl. Ed., p. 210.
133. M. Fettouhi, L. Ouhab,, A. Boukhari, O. Cador, C. Mathoniére and O. Kahn, *Inorg. Chem.*, **35**, 4932 (1996).
134. P. Halasyamani, M.J. Willis, Ch.L. Stern, P.M. Lundquist, G.K. Wong and K.R. Poeppelmeier, *Inorg. Chem.*, **35**, 1367 (1996).
135. Ii Jianmin, Ke Yanxiong, Wang Quanming and Wu Xintao, *Cryst. Res. Technol.*, **31**, 453 (1996).
136. Dong Chen, Shi-ying Han, Shi-iong Zhu, Wen-xia Tang, An-bang Dai, Ming Wang, Bo-yi Wang and Pei-ju Zheng, *Sc. China, Ser. B,* **35**, 1036 (1992).
137. G.M. Gray, A.L. Zell and Einspahr, *Inorg. Chem.*, **15**, 2923 (1986).
138. A.L. Abuhijleh and C. Woods, *J. Chem. Soc., Dalton Trans.*, 1249 (1992).
139. A.L. Abuhijleh, J.Pollitte and Woods, *Inorg Chim. Acta,* **215**, 131 (1994).
140. M. Bracci, C. Ercolani, B. Floris, M. Bassetti, A. Chien-Villa and C. Guastini, *J. Chem. Soc., Dalton Trans.*, 1357 (1990).
141. R. Weber and G. Berghoff, *Z. Kristallogr.*, **207**, 182 (1993).
142. A.J. Blake, R.D. Gould, C.M. Grant, P.E.Y. Milne, D. Reed and R.E.P. Winpenny, *Angew. Chem. Int. Ed. Engl.*, **33**, 195 (1994).
143. A.J. Blake, P.J.Y. Milne and R.E.P. Winpenny, *J. Chem. Soc., Dalton Trans.*, 3727 (1993); A.J. Blake, R.D. Gould, P.E.Y. Milne and R.E.P. Winpenny, *J. Chem. Soc., Chem. Commun.*, 522 (1992).
144. C. Benelli, A.J. Blake, P.E.Y Milne, J.M Rawson and R.E.P Winpenny, *Chem. Eur. J.*, **1**, 614 (1995).
145. Y. Kobuke, K. Kokubo and M. Munakara, *J. Am Chem. Soc.*, **117**, 12751 (1995).
146. I.A. Kahva, D. Miller, M. Mitchel, F.R. Franczek, R.G. Goodrich, D.J. Wiliams, C.A. Mahoney, A.M.Z. Slawin, S.V. Ley and Ch.J. Groombridge, *Inorg. Chem.*, **31**, 3963 (1992)
147. Xiao-Ming Chen and T.C.W. Mak, *Polyhedron,* **13**, 1087 (1994).
148. N.N Sauer, E. Garcia, K.V. Salazar, R.R.Ryan and J.A. Martin, *J. Am. Chem. Soc.*, **112**, 1524 (1990).

149. S.R. Breeze, S. Wang, J.E. Gredan and N.P.Rajer, *J. Chem. Soc., Dalton Trans.*, 2327 (1998).
150. O.Ya. Vasilyeva, V.N. Kokozay, N.A. Zhukova and L.A. Kovbasyuk, *Polyhedron,* **16**, 263 (1997).
151. K.S. Chang, S.J. Retting, A. Storr and J. Trotter, *Can. J. Chem.*, **55**, 4166 (1977).
152. S.M.Melinak, and D.Coucouvanis, *Inorg. Chem.*, **35**, 4810 (1996).
153. Yan Zang, Zhanfeng Yin, Guoxiang Wang, Cheng Zeng, Anbang Dai and Zhongyuan Zhou, *Inorg. Chem.*, **29**, 560 (1990).
154. R. Castro, M.L. Durán, J.A. Garcia-Vázquez, J. Romero, A. Sousa, A. Castiñeiras, W. Hiller and J. Strähle, *Polyhedron,* **11**, 1195 (1992).
155. D.M.L. Goodgame, D.J. Wiliams and R.E.P. Winpenny, *Polyhedron,* **8**, 873 (1989).
156. C. Benelli, A. Caneschi, D. Gateschi, O. Guillou and L. Pardi, *Inorg. Chem.*, **29**, 1750 (1990).
157. A.J. Blake, V.A. Cherepanov, A.A. Dunlop, C.M. Grant, P.E.Y. Milne. J.M. Ravson and R.E.P. Winpenny, *J.Chem. Soc., Dalton Trans.*, 2719 (1994).
158. A.N. Papadopoulos, C.P. Raptopoulou,A. Terzis, A.G. Hatzidimitriou, A. Gourdon and D.P.Kessissoglou, *J. Chem. Soc., Dalton Trans.*, 2591 (1995).
159. L. Antolini, L.P. Battaglia, A. Bonamartini-Corradi, L. Menabue, G. Micera and M. Saladini, *Inorg. Chem.*, **25**, 3301 (1986).
160. S. Wang, S.J. Trepanier and M.J. Wagner, *Inorg. Chem.*, **32**, 833 (1993).
161. W. Bidell, V. Shklover and H. Berke, *Inorg. Chem.*, **31**, 5561 (1992).
162. R. Prins, M. Biagini-Cingi, M. Drillon, R.A.G. de Graff, J. Hasnoot, A.M. Manoti-Lanfredi, P. Rabu, J. Reedijk and F. Ugozzoli, *Inorg. Chim. Acta,* **248**, 35 (1996).
163. Xiao-Ming Chen, Ming-Liang Tong, Yo-Luan Wu and Yu-Li Lao, *J. Chem Soc., Dalton Trans.*, 2181 (1996).
164. I.A. Kahwa, D. Miller, M. Mitchel and F.R. Fronczek, *Acta Cryst., Sect. C,* **49**, 320 (1993); N.S. Fender, S.S. Finegan, D. Miller, M. Mitchel, I.A. Kahwa and F.R. Fronczek, *Inorg. Chem.*, **33**, 4002 (1994).
165. A. Benito, J. Cano, R. Martinez-Máñez, J. Soto, J. Payá, F. Lloret, M. Julve, J. Faus and M.D. Marcos, *Inorg. Chem.*, **32**, 1197 (1993).
166. M. Mizutani, K. Jitsukawa, H. Matsuda and H. Einaga, *J. Chem. Soc., Chem. Commun.*, 1389 (1996).
167. A.G. Stemmler, A. Barwiski, M.J. Baldwin, V. Young and V.L. Pecoraro, *J. Am. Chem. Soc.,* **118**, 11962 (1996).
168. M. Andruch, I. Ramade, E. Codjovi, O. Guillou, O. Kahn and J.Ch. Trombe, *J. Am. Chem. Soc.,* **115**, 1822 (1993).
169. M.R. Churchil, Yong-Ji Li, D. Nalewajek, P.M. Scraber and J. Dorfman, *Inorg. Chem.*, **24**, 2684 (1985).
170. I.N. Polyakova, V.S. Sergienko, A.L. Poznyak and O.G. Elbert, *Zh. Neorg. Khim.*, **42**, 1653 (1997); Engl. Ed., p. 1509.
171. R.T. Baker, S.J. Rettig, A. Storr and J. Trotter, *Can. J. Chem.*, **54**, 343 (1976).
172. R.J. Parker, D.C.R. Hockless, B. Moubaraki, K.S. Murray and L. Spiccia, *J. Chem. Soc., Chem. Commun.*, 2789 (1996).
173. J.A. Samuels, Wen-C. Chiang, J.C. Huffman, K.L. Trojan, W.E. Hatfield, D.V. Baxter and K.G. Caulton, *Inorg. Chem.*, **33**, 2167 (1994).

174. Suning Wang, *Inorg. Chem.*, **30**, 2252 (1991).
175. M.A. Bañares, L. Dauphin, Xinjian Lei, Wei Cen, Maoyu Shang, E.E. Wolf and Th.P. Fehlner, *Chem. Mater.*, **7**, 553 (1995).
176. Jing-Yang Niu, Jing-Ping Wang, Xiao-Zeng You, T.C.W. Mak and Zhong-yuan Zhou, *Polyhedron*, **15**, 3963 (1996).
177. M.A. Pervuchina, V.S. Sergienko, G. G. Sadikov, I.V. Tatyanina and E.A. Torchenskova, *Koord. Khim.*, **16**, 354 (1990); Engl. Ed. p. 195.
178. Xiao-Ming Chen, Yu-Luan Wu, Ye-Xiang Tong and Xiao-Ying Huang, *J. Chem. Soc., Dalton Trans.*, 2443 (1996).
179. M. Biagini Cingi, A.M. Manoti Lanfredi, A. Tiripicchio and M. Tiripicchio Camelini, *Acta Cryst.,Sect.B*, **33**, 659 (1977).
180. M. Klaassen and P. Klüfers, *Z. Anorg. Allg. Chem.*, **619**, 661 (1993).
181. M. Klaassen and P. Klüfers, *Acta Cryst., Sect. C*, **50**, 686 (1994).
182. N. Habermann, G. Jung, M. Klaassen and P. Klüfers, *Chem. Ber.*, **125**, 809 (1992).
183. R. Fuchs, N. Habermann and P. Klüfers, *Angew. Chem., Int. Ed. Engl.*, **32**, 852 (1993).
184. M. Biagini Cingi, A.M. Manoti Lanfredi, A. Tiripicchio and M. Tiripicchio Camelini, *Acta Cryst.,Sect.B*, **34**, 412 (1978).
185. V.A. Igonin, S.V. Lindeman, Yu.T. Struchkov, Yu.A. Molodtsova, O.I. Shchegolikhina and A.A. Zhdanov, *Izv. Akad. Nauk, Ser. Khim.*, 752 (1993); Engl. Ed. p. 718.
186. J. Valo and M. Näsäkkälä, *Acta Chem. Scand.*, **48**, 20 (1994).
187. M. Biagini Cingi, A.M. Manoti Lanfredi, A. Tiripicchio and M. Tiripicchio Camelini, *Acta Cryst.,Sect.B*, **34**, 406 (1978).
188. V. Bangera Kasthuri, P. Mohan Rao and M. Nethaji, *Cryst. Res. Technol.*, **31**, 287 (1996).
189. F. Sapiña, E. Escrivá, J.V. Folgado, A. Beltrán, D. Beltrán, A. Fuertesand M. Drillon, *Inorg. Chem.*, **31**, 3851 (1992).
190. I. Sevilla, R. Ortiz, L. Perelló, J. Borrás, S. Garcia-Granda and A. Aguirne, *J. Inorg. Biochem.*, **57**, 147 (1995).
191. Ö. Berkaroglu, M. El Sharif, H. Endres and H.J. Keller, *Acta Cryst., Sect B*, **32**, 2983 (1976).
192. A. Gleizes and M. Verdaguer, *J. Am. Chem. Soc.*, **103**, 7373 (1981).
193. Suning Wang, S.J. Trepanier, Jian-Chuang Zheng, Zheng Pang and M.J. Wagner, *Inorg. Chem.*, **31**, 2118 (1992).
194. H. Milburn, M.R. Truter and B.L. Vickery, *J. Chem. Soc., Dalton Trans.*, 841 (1974).
195. R.D. Hart, B.W. Skelton and A.H. White, *Aust. J. Chem.*,**45**, 1927 (1992).
196. S. Kotila and J. Valkonen, *Acta Chem. Scand.*, **48**, 312 (1994).
197. J.R.D. DeBord, Yiping Zhang, R.C. Haushalter, J. Zubieta and Ch.J. O'Connor, *J. Solid. State., Chem.*, **122**, 251 (1996).
198. Yu Pei, K. Nakatani, O. Kahn, J. Sletten and J.P. Renard, *Inorg. Chem.*, **28**, 3170 (1989).
199. F. Lloret, M. Julve, R. Ruiz, Y. Journaux, K. Nakatani, O. Kahn and J. Sletten, *Inorg. Chem.*, **32**, 27 (1993).
200. I. Morgenstern-Badarau, D. Laroque, E. Bill, H. Winkler, A.X. Trautwein, F. Robert and Y. Jeannin, *Inorg. Chem.*, **30**, 3180 (1991).

201. H.O. Stumpf, Yu Pei, L. Ouahab, F. Le Berre, E. Codjovi and O. Kahn, *Inorg. Chem.*, **32**, 5687 (1993).
202. H.O. Stumpf, L. Ouahab, Yu Pei, D. Gandjean and O. Kahn, *Science,* **261**, 447 (1993).
203. R.L. Oushoorn, K. Boubekeur, P. Batail, O. Guillou and O. Kahn, *Bull. Soc. Chim. Fr.*, **133**, 777 (1996).
204. G. Smith, D.S. Sagatys, R.C. Bott, D.E. Lynch and C.H.L. Kennard, *Polyhedron,* **11**, 631 (1992).
205. M. Biagini Cingi, A.M. Manoti Lanfredi, A. Tiripicchio and M. Tiripicchio Camelini, *Acta Cryst.,Sect.B,* **35**, 312 (1979).
206. O. Guillou, P. Bergerat, O. Kahn, E. Bakalbassis, K. Boubekeur, P. Batail and M. Guillot, *Inorg. Chem.*, **31**, 110 (1992).
207. C. C. Bazzicalupi, A. Bencini, A. Bianchi, V. Fusi, P. Paoletti., G. Picardi and B. Valtancoli, *Inorg. Chem.*, **34**, 5622 (1995).
208. Huang Jinling, Lu Jiaxi, Huang Debin and Liu Shixiong, *Sc. Sinica, Ser.B,* **26**, 19 (1983).
209. G.O. Morpurgo, V. Mosini, P. Porta, G. Dessy and V. Fares, *J. Chem. Soc., Dalton Trans.*, 1272 (1980).
210. G.O. Morpurgo, V. Mosini, P. Porta, G. Dessy and V. Fares, *J. Chem. Soc., Dalton Trans.*, 111 (1981).
211. V.V. Fomenko, L.I. Kanaeva, M.A. Porai-Koshits and T.N. Polynova, *Zh. Strukt. Khim.*, **15**, 268 (1974).
212. D.P.Kessissoglou, C.P. Raptopoulou, E.G. Bakalbassis, A. Terzis and J. Mrozinski, *Inorg. Chem.*, **31**, 4339, (1992).
213. Mei Fang Wu, Zhong Wing Chen, Jun Qiu , We Xia Tang, Jing Ling Wang and Fang Ming Miao, *Chinese Chem. Let.*, **5**, 713 (1994) ; Zhong Ning Chen, Jing Ling Wang, Jun Qiu, Fang Ming Miao and Wen Xia Tang, *Inorg. Chem.*, **34**, 2255 (1995).
214. L. Antolini, L.P. Bataglia, G. Battistuzzi Gavioli, A. Bonamartini Coradi, G. Grandi, G. Marcotrigiano, L. Menabue and G.C. Pellacani, *J. Am. Chem. Soc.*, **105**, 4327 (1983).
215. Yu Pei, O. Kahn and J. Sletten, *J. Am. Chem. Soc.,* **108**, 3141 (1986); Yu Pei, M. Verdaguer, O. Kahn, J. Sletten and J.P. Renard, *Inorg. Chem.*, **26**, 138 (1987).
216. Yu Pei, M. Verdaguer, O. Kahn, J. Sletten and J.P. Renard, *J. Am. Chem. Soc.*, **108**, 7428 (1986).
217. V. Baron, B. Gillon, J. Sletten, C. Mathoniere, E. Codjovi and O. Kahn, *Inor. Chim. Acta,* **235**, 69 (1995).
218. H.O. Stumpf, Yu Pei, O. Kahn, J. Sletten and J.P. Renard, *J. Am. Chem. Soc.*, **115**, 6738 (1993).
219. Yu Pei, O. Kahn, J. Sletten, J.P. Renard, R. Georges, J.C. Gianduzzo, J. Curely and Qiang Xu, *Inorg. Chem.*, **27**, 47 (1988).
220. D.W. Harrison and J.C.G. Bünzli, *Inorg., Chim. Acta,* **109**, 185 (1985).
221. O. Guillou, R.L. Oushoorn, O. Kahn, K. Boubekeur and P. Batail, *Angew. Chem.*, **104**, 658 (1992) ; *Angew. Chem. Int. Ed. Engl.*, **31**, 626 (1992).
222. O. Guillou, O. Kahn, R.L. Oushoorn, K. Boubeker and P. Batail, *Inorg. Chim. Acta,* **198-200**, 119 (1992).
223. M.A.S. Goher and F.A. Mautner, *Polyhedron,* **13**, 2149 (1994).

224. Yu. T. Struchkov, Yu.N. Belokon, V.M. Belikov, I.E. Zal'tser and G.G. Alexandrov, *J. Organomet. Chem.*, **210**, 411 (1981).
225. D. Schulz, Th. Weyhermüller, K. Wieghardt, Ch. Butzlaff and A.X. Trautwein, *Inorg. Chim. Acta,* **246**, 387 (1996).
226. V. Kettmann, J. Krätsmár-Šmogroviè, O. Švajlenová and M. Žemlička, *Z. Naturforsch.* **47b**, 1565 (1992).
227. J.L. Atwood, G.V.Orr, N.C. Means, F. Hamada, H. Zhang, S.G. Bott and K.D. Robinson, *Inorg. Chem.*, **31**, 603 (1992).
228. A. Gleizes, F. Maury and J. Galy, *Inorg. Chem.*, **19**, 2074 (1980).
229. P. Chananont, P.E. Nikan, J.M. Waters and T.N. Waters, *Acta Cryst., Sect. B,* **36**, 2145 (1980).
230. T. Weichert and J. Löhn, *Z. Kristallogr.*, **139**, 223 (1974).
231. W. Pannhorst and J. Löhn, *Z. Kristallogr.*, **139**, 236 (1974).
232. M.J. Sanchis, P. Gomez-Romero, J.V. Folgado, F. Sapiña, R. Ibáňey, A. Beltrán, J. Garcia and D. Beltrán, *Inorg. Chem.*, **31**, 2915 (1992).
233. G. Polla, G. Leyva, P.K. de Perazzo, H. Lanza and M.A.R. de Benyacar, *J. Solid. State., Chem.*, **117**, 145 (1995).
234. R.F. Baggio, P.K. de Perazo and G. Polla, *Acta Cryst.,, Sect. C,* **41**, 194 (1985).
235. M. Isausti, M.K. Urtiaga, R. Cortés, J.L. Mesa, M.J.Arriortua and T. Rojo, *J. Mater. Chem.*, **4**, 1867 (1994).
236. T. Ogata, T. Taga and K. Osaki, *Bull. Chem. Soc. Jpn.*, **50**, 1680 (1977).
237. W. Bensch, N. Seferiadis and H.R. Oswald, *Inorg. Chim. Acta,* **126**, 113 (1987).
238. J. Kim, J.M. Lim, Yong-Kook Choi and Youngkyu Do, *Angew. Chem. Int. Ed. Engl.*, **35**, 998 (1996).
239. Jiang-Gao Mao, Ling Song, Xiao-Ying Huang and Jin-Shun Huang, *Polyhedron,* **16**, 963 (1997).
240. Shi Jing-Min, Yan Shi-Ping, Liao Dai-Zheng, Jiang Zong-Hui, Wang Geng-Lin, Wang Ri-Ji, Wang Hong-Gen and Yao Xin-Kan, *Chinese J. Struct. Chem., (Jiegou Huaxue),* **14**, 102 (1995).
241. A. Bouayad, Ch. Brouca-Cabarrecq, J.Ch. Trombe and A. Gleizes, *Inorg. Chim. Acta,* **195**, 193 (1992).
242. H. Scooloudi, *Acta Cryst.*, **6**, 651 (1953).
243. J. Černák, F. Gérard and J. Chomič, *Acta Cryst., Sect. C,* **49**, 1294 (1993).
244. H.Yuge and T. Iwamoto, *J. Chem. Soc., Dalton Trans.*, 1237 (1994).
245. T. Miyoshi, T. Iwamoto and Y. Sasaki, *Inorg. Chim. Acta,* **7**, 97 (1973).
246. S.H. Whitlow, *Inorg. Chem.*, **12**, 2286 (1973)
247. V.V. Fomenko, T.N. Polynova and M. Porai-Koshits, *Zh. Strukt. Khim.,* **16**, 645 (1975); Engl. Ed., p. 597.
248. B.F. Mullica, D.B. Tippin and E.L. Sappenfield, *J. Coord. Chem.*, **25**, 175 (1992).
249. X. Solans, M. Font-Altaba, J. Oliva and J. Herrera, *Acta Cryst., Sect. C,* **39**, 435 (1983).
250. Ch. Brouca-Cabarrecq, B. Marot and A. Moset, *Acta Cryst., Sect. C,* **52**, 1903 (1996).
251. S. Kotila and J. Valkonen, *Acta Chem. Scand.*, **47**, 957 (1993).
252. N. Nakasuka, Sh. Azuma and M. Tanaka, *Acta Cryst., Sect. C,* **42**, 1482 (1986).
253. J. Soldánová, F. Pavelčík and J. Majer, *Acta Cryst., Sect. B,* **37**, 921 (1981).

254. Ya. M. Nesterova, M. A. Porai-Koshits and V.A. Logvinenko, *Zh. Neorg. Khim.*, **24**, 2273 (1979); Engl. Ed., p. 1262; Ya. M. Nesterova and M.A. Porai-Koshits, *Koord. Khim.*, **10**, 129 (1984).
255. M. Inasausti, J.L. Pizarro, L. Lezama, R. Cortes, E.H. Bocanegra, M.I. Arriortua and T. Rojo, *Chem. Mater.*, **6**, 707 (1994).
256. J.L. Atwood, G.W. Orr, F. Hamada, S.G. Bott and K.D. Robinson, *Supramol. Chem.*, **1**, 15 (1992).
257. H.O. Stumpf, L. Ouhab, Yu Pei, P. Bergerat and O. Kahn, *J. Am. Chem. Soc.*, **116**, 3866 (1994).
258. T. Hosokawa, M. Takano, S.I. Murahashi, H. Ozaki, Y. Kitagawa, K. Sakaguchi and Y. Katsube, *J. Chem. Soc., Chem. Commun.*, 1433 (1994).
259. S. Halasyamani, K.R. Heier, Ch.L. Stern and K.P. Poeppelmeier, *Acta Cryst., Sect. C,* **53**, 1240 (1997).
260. T.C. Hodashova, M.A. Porai-Koshits, N.K. Davidenko and N.N. Vlasova, *Koord. Khim.*, **10**, 262 (1984).
261. A.J. Blake, I.A. Fallis, R.O. Gould, S. Parsons, S.A. Ross and M. Schröder, *J. Chem. Soc., Chem. Commun.*, 2467 (1994).
262. Yu.A. Simonov, V. Kh. Kravtsov, I.O. Fritskii, E.E. Gubina, R.D. Lampeka, T.S. Iskenderov and A.Zh. Zhumabaev, *Koord. Khim.*, **21**, 407 (1995); Engl. Ed., p. 389.
263. G.B. Jameson, N. Seferiadis and H.R. Oswald, *Acta Cryst., Sect. C,* **42**, 984 (1986).
264. P. Gomez-Romero, G.B. Jameson, J.J. Borras-Almenar, E. Escriva, E. Coronado and D. Beltrán, *J. Chem. Soc., Dalton Trans.*, 2747 (1988).
265. H. Oshio, K. Toriumi, S. Bandow, K. Miyagawa and S. Kurita, *J. Chem. Soc., Dalton Trans.*, 10134 (1990).
266. A. Korczynski, *Rocz. Chem.*, **36**, 1539 (1962).
267. T. Hosokawa, M. Takano and S. Murahashi, *J. Am. Chem. Soc.*, **118**, 3990 (1996).
268. R. Ősterberg, B. Sjöberg and R. Söderquist, *Acta Chem. Scand.*, **26**, 4184 (1972).
269. Li-Feng Mao and A. Mayr, *Inorg. Chem.*, **35**, 3183 (1996).
270. Gou Shao-Hua, Xu Zheng, You Xiao-Zeng, Zheng Pei-Ju and Dong Jun, *Chinese J. Chem.*, **10**, 232 (1992).
271. E. Papavinasam, *Z. Kristallogr.*, **197**, 217 (1991).
272. Xiao-Ming Chen and T.C.W. Mak, *Inorg, Chem.*, **33**, 2444 (1994).
273. D.A. Langs and C.R. Hare, *J. Chem. Soc., Chem. Commun.*, 890 (1967).
274. E.A. Klop, A.J.M. Duisenberg and A.L. Spek, *Acta Cryst., Sect. C,* **39**, 1342 (1983).

Table 1. Crystallographic and structural data for hetero - double salts of copper(II)[a].

COMPOUND (colour)	CRYST. CL. SPACE G. Z	a[Å] b[Å] c[Å]	α[°] β[°] γ[°]	CHROMO PHORE	M - L [Å]		L - M - L [°]		REF.
$[Cu(C_2H_4O_2)_2][Ba(C_2H_6O_2)_6]$ (deep blue) (at 203 K)	m Cc 4	12.103(6) 13.527(3) 17.091(8)	 93.17(2)	$Cu^{II}O_4$	O[b]	1.922(4,12)	O,O[b]	86.7(2,2)[c] 93.3(2,4)	10
				$Ba^{II}O_9$	O	2.795(5,34) 2.907(7,6)	O,O	not given	
$[Cu(NH_3)_4][PtCl_4]$ (deep purple)	tg P4/mnc	9.036(3) 6.441(2)		$Cu^{II}N_4$	N	2.00(1,0)	N,N	not given	11
				$Pt^{II}Cl_4$	Cl	2.298(4,0)	Cl,Cl	not given	
$[Cu(en)_2][PtCl_2(en)_2](ClO_4)_4$ (red)	or Icma 2	13.645(1) 10.787(1) 9.645(1)		$Cu^{II}N_4$	enN	2.031(4)	N,N	91.2(1)	12
				$Pt^{IV}N_4Cl_2$	enN Cl	2.031(4) 2.313(3)	N,N	91.2(1)	
$[Cu(en)_2][PtBr_2(en)_2](ClO_4)_4$ (not given)	or Icma 2	13.599(1) 10.917(1) 9.657(1)		$Cu^{II}N_4$	enN	2.043(3)	N,Br	90.9(1)	12
				$Pt^{IV}N_4Br_2$	enN Br	2.043(3) 2.453(1)	N,Br	90.9(1)	
$[Cu(en)_2][PtBr_2(en)_2](ClO_4)_4$ (not given)	or Icma 2	13.625(2) 10.905(2) 9.651(1)		$Cu^{II}N_4$	enN	2.043(6)	N,Br	90.9(2)	12
				$Pt^{IV}N_4Br_2$	enN Br	2.043(6) 2.436(2)	N,Br	90.9(2)	
$[Cu(papH)_4][Li(H_2O)_4].2H_2O$ (orange)	m $P2_1/n$ 4	7.1790(10) 27.647(6) 8.824(2)	 94.50(3)	$Cu^{II}N_4$	N	1.918(4,8) 1.958(4,4)	N,N	82.2(2,6)[c] 98.6(2)[d] 96.8(2) 175.3(2,1.5)	13a
				$Li^{I}O_4$	O	not given	O,O	not given	

Table 1. Continued 2.

$[Cu(oep)]_2[SbCl_6].0.5CH_2Cl_2.$	m	14.791(3)		$Cu^{II}N_4$	N	1.984(6)	N,N	not given	13b
$0.5CHCl_3$	$P2_1/c$	17.042(5)	99.87(2)						
(not given)	2	14.796(3)		$Sb^{V}Cl_6$	Cl	not given	Cl,Cl	not given	
(at 124 K)									
$[CuCl_4][Pt(NH_3)_4]$	m	7.687(2)		$Cu^{II}Cl_4$	Cl	2.287(4,16)	Cl,Cl	90.3(1)	11
(green)	$P2_1/c$	7.941(2)	91.61(1)						
		8.057(2)		$Pt^{II}N_4$	N	2.04(1,1)	N,N	89.9(5)	
							N,N'	86.5(1,6)	
$[CuCl_4][Pt(en)_2Cl_2].H_2O$	m	7.788(6)		$Cu^{II}Cl_4$	Cl	2.239(4,0)	Cl,Cl	93.0(3,4.1)	14
(grass-green)	$P2_1/c$	7.369(7)	114.4(1)			2.276(5,0)		154.1(2)	
	2	15.080(20)							
				$Pt^{IV}N_4Cl_2$	enN	2.081(12,7)	N,N	82.7(5,0)[c]	
					Cl	2.313(4,0)	N,Cl	88.9(4,6)	
$[CuCl_4][Sc(15\text{-crown-}5)Cl_2]_2$	m	10.475(3)		$Cu^{II}Cl_4$	Cl	2.236(4,0)	Cl,Cl	100.6(2,6)	15
(brown)	$P2_1/m$	18.895(4)				2.260(7,10)		129.3(2,2.0)	
	2	10.170(2)	115.30(2)						
				$Sc^{III}O_5Cl_2$	O	2.11(2,2)	O,O	72.9(12,5)	
					Cl	2.422(6,5)		141.1(12)	
							Cl,Cl	179.8(4)	
							O,Cl	89.6(5,3.4)	
				$Sc^{III}O_5Cl_2$	O	2.15(2,1)	O,O	72.5(5,1.1)	
					Cl	2.430(8,5)		141.7(6)	
							Cl,Cl	176.4(3)	
							O,Cl	89.2(5,1.5)	
$[CuCl_4][Ba(15\text{-crown-}5)_2]$	m	12.119(2)		$Cu^{II}Cl_4$	Cl	2.275(5,7)	Cl,Cl	99.7(2,2.9)	15
(yellow)	Pn	9.386(2)	93.81(2)					131.3(2,4)	
	2	13.475(3)							
				$Ba^{II}O_{10}$	O	2.86(1,6)	O,O	58.5(2)	
								172.3(3)	
$[CuCl_4][Mg(15\text{-crown-}5)(H_2O)_2].$	or	9.525(3)		$Cu^{II}Cl_4$	Cl	not given	Cl,Cl	not given	16
H_2O	Ccc2	12.535(3)							
(dark brown)	4	16.520(4)		$Mg^{II}O_7$	O	not given	O,O	not given	
					H_2O	not given			

Table 1. Continued 3.

$[CuCl_4][Co(15\text{-crown-}5)(H_2O)_2]$ (dark brown)	or Ccc2 4	9.505(2) 12.431(3) 16.454(4)		$Cu^{II}Cl_4$	Cl	2.225(9,1) 2.277(9,0)	Cl,Cl	97.3(6,3.1) 140.8(4,0)	17a
				$Co^{II}O_7$	O H_2O	2.14(1,1) 2.23(1) 2.09(1,0)	O,O	72.7(7,1.0) 85.7(6,5) 96.2(6,3.6) 142.9(7,6) 173.2(12)	
$[Cl_2Cu(\mu\text{-}Cl)_2CuCl_2][Mg(15\text{-}$ $\text{crown-}5)(MeCN)_2]$ (yellow orange)	tr P-1 1	7.248(2) 9.542(3) 9.470(2)	82.30(2) 97.04(2) 93.90(2)	$Cu^{II}Cl_4$	Cl μCl	2.19(1) 2.31(-,2)	Cl,Cl	86.6 - 144	16
				$Mg^{II}O_5N_2$	O N	2.121(7,29) 2.220(7,10) 2.158(4)	O,O O,N	72.4(2,1.4)[c] 89.8(2,4.8)	
$[Cl_2Cu(\mu\text{-}Cl)_2CuCl_2][Co(15\text{-}$ $\text{crown-}5)(MeCN)_2]$ (brown orange)	tr P-1 1	7.243(2) 9.383(3) 9.566(3)	82.32(2) 86.13(2) 83.13(2)	$Cu^{II}Cl_4$	Cl μCl	2.193(3,12) 2.315(3,22)	Cl,Cl	94.4(1,7.8) 102.2(1) 144.3(1)	17a
				$Co^{II}O_3N_2$	O N	2.153(14,50) 2.255(9,15) 2.0968(8)	N,O O,O	85.7(4,1.9) 94.4(4,2.1) 36.6(5,1.4) 72.3(5,2) 107.5(3,3) 142.9(3)	
$[Cl_2Cu(\mu\text{-}Cl)_2CuCl_2][K(n\text{-}15\text{-}$ $\text{crown-}5)_2]_2.3H_2O$ (brown-green)	tr P-1 2	24.311(8) 15.314(4) 13.124(3)	67.63(2) 75.78(2) 95.44(2)	$Cu^{II}Cl_4$	Cl μCl	2.167(6,10) 2.332(6,20)	Cl,Cl	88.7(2,2) 103.8(2,1.7) 129.7(2,5.2)	17b
				$K^{I}O_{10}$	O	2.818(9,61) 2.960(9,136)	O,O	not given	
$[Cu(mnt)_2][Fe(cp^*)_2]_2$ (black)	tr P-1 1	9.713(5) 11.407(4) 11.958(5)	100.90(2) 113.20(5) 92.66(3)	$Cu^{II}S_4$	S	2.262(2,0) 2.279(1,0)	S,S	90.67(5)	18
				FeC_{10}	cp*C	not given	C,C	not given	

Table 1. Continued 4.

$[Cu(D\text{-}lysine)(L\text{-}lysine)][Hg_2I_6]$	m	21.863(2)		$Cu^{II}O_2N_2$	O	1.92(1)		not given	19
(not given)	C2/c	9.958(1)	110.60(3)		N	1.99(2)			
	4	15.402(2)							
				$Hg^{II}I_4$	I	not given		not given	
$[Cu(L\text{-}lysine)_2][Hg_2I_6]$	m	22.099(2)		$Cu^{II}O_2N_2$	O	1.96(2)		not given	19
(not given)	C2	9.897(1)	111.82(2)		N	1.99(4)			
	4	15.648(2)							
				$Hg^{II}I_4$	I	not given		not given	
$[Cu(L\text{-}lysine)_2][Hg_2I_6].H_2O$	m	22.630(2)		$Cu^{II}O_2N_2$	O	1.94(2)		not given	19
(not given)	C2	10.056(1)	111.74(3)		N	1.99(4)			
	4	16.125(2)							
				$Hg^{II}I_4$	I	not given		not given	
$[Cu(hdbc)][Fe(\eta^5\text{-}cp)(CO)_2(NH_3)]$	tr	9.850(1)	103.97(1)	$Cu^{II}O_2N_2$	O	1.896(5,3)	O,O	88.9(2)	20
$.PF_6$	P1	9.866(1)	99.16(1)		N	1.936(5,1)	N,N	84.7(2)[c]	
(yellow)	2	18.176(2)	89.66(1)				O,N	93.7(2,1)[d]	
								172.2(2,1)	
				FeC_7N	cpC	2.10(1,2)	C,C	92.8(3)	
					OC	1.780(7,9)	C,N	94.2(3,7)	
					H_3N	2.008(6)			
$[CuCl_5][Co(NH_3)_6]$	c	21.992(3)		$Cu^{II}Cl_5$	Cl_{eq}	2.392(2,0)	Cl,Cl	not given	21
(not given)	Fd3c				Cl_{ax}	2.291(2,0)			
	32								
				$Co^{III}N_6$	N	1.968(2,0)	N,N	90.0(4,3)	
$[Cl_3Cu(\mu\text{-}Cl)_2CuCl_3][Co(en)_3]_2Cl_2$	or	13.560(9)		$Cu^{II}Cl_5$	Cl	2.268(5,6)	Cl,Cl	90.5(2,5.8)	22
$.2H_2O$	Pbca	14.569(9)				2.319(5)		118.3(2)	
(orange)	8	17.885(12)			μCl	2.325(4)		145.3(2)	
						2.703(5)		172.6(2)	
				$Co^{III}N_6$	enN	1.963(13,10)	N,N	83.7(5,7)[c]	
						1.998(12)		91.9(5,2.1)	
								175.0(6,3)	

Table 1. Continued 5.

$[Cl_3Cu(\mu\text{-}Cl)_2CuCl_3][Ir(en)_3]_2Cl_2$	or	13.705(5)		$Cu^{II}Cl_5$	Cl	2.271(2,18)	Cl,Cl	90.6(1,5.4)	23
$.2H_2O$	Pbca	14.467(19)				2.321(2)		117.10(8)	
(orange)	4	18.091(6)			μCl	2.305(2)		146.37(9)	
						2.705(2)		174.27(9)	
				$Ir^{III}N_6$	enN	2.080(7,17)	N,N	82.1(2,1)	
								92.7(2,2.9)	
								174.0(3,3)	
$[Cl_3Cu(\mu\text{-}Cl)_2CuCl_3][Rh(en)_3]_2Cl_2$	or	13.639(6)		$Cu^{II}Cl_5$	Cl	2.269(1,10)	Cl,Cl	90.5(1,5.1)	23
$.2H_2O$	Pbca	14.577(11)				2.315(1)		117.18(3)	
(orange)	4	18.020(6)			μCl	2.318(1)		146.56(4)	
						2.705(1)		173.88(4)	
				$Rh^{III}N_6$	enN	2.068(2,19)	N,N	82.6(1,3)[c]	
								92.5(1,2.7)	
								174.0(1,8)	
$[Cu(trisH)(tris)(H_2O)]_2CrO_4$[c]	tr	6.424(1)	98.60(1)	$Cu^{II}O_3N_2$	O	1.957(3,30)	O,O	98.3(1,5.8)	24
(green)	P-1	10.735(1)	89.97(1)		N	2.000(3,9)	N,N	96.1(1)	
	2	21.578(2)	72.63(1)		H_2O	2.317(3)	O,N	84.8(1,1.7)[c]	
								168.6(1,4.3)	
				$Cr^{VI}O_4$	O	1.636(8,71)	O,O	109.5(4,3.2)	
				$Cu^{II}O_3N_2$	O	1.956(2,28)	O,O	98.3(1,5.8)	
					N	2.007(3,10)	N,N	96.4(1)	
					H_2O	2.297(3)	O,N	84.8(1,1.7)[c]	
								168.7(1,4.3)	
				$Cr^{VI}O_4$	O	1.633(8,63)	O,O	109.5(4,3.0)	

Table 1. Continued 6.

Compound	Cryst. cl. / Space gr. / Z	a, b, c	α, β, γ	Chromophore	L	M–L	L,L	L–M–L	Ref.
[((bpac)Cu)$_2$(μ-O)$_2$](SbF$_6$)$_2$	m	26.1916(6)		Cu^{II}N$_3$O$_2$	N	1.989(-,1)	N,N	99.0(2,3.4)	25
.7Me$_2$CO.2MeCN	C2/c	15.3560(4)	96.493(1)			2.298(-)		171.1(2)	
(brown)	8	22.2705(6)			μO	1.806(-,3)	O,O	78.6(2)	
							O,N	100.0(2,3.8)	
								171.7(2,6)	
				Sb^{V}F$_6$	F	not given	F,F	not given	
[Cu(H$_2$O)$_6$]$_2$[ZrF$_8$]	m	15.924(15)		Cu^{II}O$_6$	H$_2$O	1.939(10,3)	O,O	86.0(4,1.4)	26
(blue)	C2/c	8.625(9)	121.69(2)			2.011(9,16)		94.0(5,3.4)	
	4	12.153(12)				2.407(13,17)		174.8(5,2.3)	
				Zr^{IV}F$_8$	F	2.103(8,36)	F,F	57.1(2,5.2)	
[Cu(μ-ddn)$_3$][SbCl$_6$]$_2$	tr	8.181(1)	112.54(8)	Cu^{II}N$_6$	N	1.989(5,0)	N,N	84.9(2)	27
(not given)	P-1	13.147(1)	99.84(7)			2.011(5,0)		91.7(2,1.2)	
(at 100 K)	1	14.314(2)	101.24(6)			2.331(5,0)			
				Sb^{V}Cl$_6$	Cl	2.356(2,9)	Cl,Cl	90.00	
						2.383(2,2)		178.50	
[Cu(tact)(MeCN)$_2$][Ni(dmit)$_2$]$_2$	tr	11.817(4)	115.59(2)	Cu^{II}N$_6$	tactN	2.023(7,5)	N,N	85.8(3)[c]	28
(black)	P-1	12.010(4)	97.72(3)		MeCN	2.491(6,0)		94.2(3)[d]	
	1	9.561(4)	95.62(3)					90.0(2,2.3)	
				Ni^{II}S$_4$	S	2.158(2,5)	S,S	93.4(1,2)[c]	
								86.6(1,2)	
								178.0(1,1.2)	

Table 1. Continued 7.

Compound	Cryst. cl. / Space gr. / Z	a, b, c	α, β, γ	Chromophore	L	M–L	L,L	L–M–L	Ref.
$[Cu_2(\mu\text{-bipym})_2(H_2O)(SO_4)_3][VO$	m	9.944(3)		$Cu^{II}N_4O_2$	N	2.045(5,49)	N,N	74.1(2,1.7)[c]	29a
$(H_2O)_5].7H_2O$	$P2_1/c$	13.928(6)	99.59(2)			2.304(5)		94.6(3,1.3)	
(blue green)	4	26.275(6)			O_3SO	2.001(4)		171.3(7)	
						2.270(4)	O,O	84.5(7)	
							N,O	92.7(2,1.2)	
								170.5(2,7)	
				$Cu^{II}N_4O_2$	N	2.002(5,1)	N,N	79.6(2,1.4)[c]	
						2.107(5)		95.4(2,1.3)	
						2.286(5)		172.3(2)	
					O_3SO	2.015(4)	O,O	84.0(2)	
					H_2O	2.246(5)	N,O	88.8(2,8)	
								94.0(2,2.3)	
				VO_6	O	1.574(5)	O,O	90.0(2,5.0)	
					H_2O	2.025(5,34)		162.0(2,5.7)	
						2.176(5)			
$[Cu(dpe)_2][(AgI)_2(\mu\text{-I})_2]$	tr	8.5402(14)	110.221(10)	$Cu^{II}N_4O_2$	N	2.010(4,10)	N,N	90.0(1,3.10)	29b
(pale brown)	P-1	10.0286(16)	95.476(10)		O	2.393(3,0)		180.0	
	1	10.584(2)	93.420(12)				O,O	180.0	
							N,O	74.2(1,1.3)[c]	
								105.4(1,1.3)	
				$Ag^{I}I_4$	I	2.790(1,10)	I,I	not given	
					μI	2.679(1,10)			
$[(tact)Cu(\mu\text{-Cl})]_2[CdCl_4]$	tg	17.841(11)		$Cu^{II}N_4Cl_2$	N	2.019(6,13)	N,N	85.3(4,1)[c]	30
(red violet)	I-42d				μCl	2.754(2,0)		94.8(2)	
	4	11.011(8)						178.9(3)	
							Cl,Cl	178.86(6)	
							N,Cl	86.3(2,1.0)	
				$Cd^{II}Cl_4$	Cl	2.467(2,0)	Cl,Cl	109.5(1,2.9)	

Table 1. Continued 8.

$[Cl_2Cu(\mu\text{-}Cl)_2Cu(H_2O)_2(\mu\text{-}Cl)_2$	tr	11.167(3)	82.43(3)	$Cu^{II}Cl_4O_2$	µCl	2.341(3)	Cl,Cl	83.9(3,4.7)	16
$CuCl_2][Mg(H_2O)_6](15\text{-crown-}5)_2$	P-1	12.242(4)	99.42(2)			2.870(3)		100.9(1)	
$.H_2O$	1	8.432(2)	84.00(2)		H_2O	1.950(9)	Cl,O	90.5(3,1.0)	
(dark brown)									
				$Cu^{II}Cl_4$	µCl	2.285(3,36)	Cl,Cl	96.9(1,3.8)	
					Cl	2.216(4,9)		139.4(1,1.7)	
				$Mg^{II}O_6$	H_2O	2.057(10,37)	O,O	90.0(4,8)	
$[Cl_2Cu(\mu\text{-}Cl)_3Cu(\mu\text{-}Cl)_3CuCl_2]$	tr	9.164(2)	122.93(2)	$Cu^{II}Cl_6$	µCl	2.249(3)	Cl,Cl	90.0(2,5.0)	16
$[Mg(15\text{-crown-}5)(MeCN)_2]$	P-1	9.900(2)	94.64(2)			2.721(4,11)		71.7(1)	
(red brown)	1	9.708(2)	97.87(2)					108.3(2)	
				$Cu^{II}Cl_5$	µCl	2.262(6,4)	Cl,Cl	89.0(2,3.6)	
						2.606(4)		103.1(1,8)	
					Cl	2.270(4)		168.7(2,1.8)	
				$Mg^{II}O_5N_2$	O	2.03(6,3)	N,O	89.9(7,6.3)	
						2.16(6)	O,O	72.0(8,6.4)[c]	
						2.37(3,4)			
					MeCN	2.14(1)			

a Where more than one chemically equivalent distance or angle is present the mean value is specified. The first number in parentheses is the e.s.d., and the second is the maximum deviation from mean value.
b The coordinated atom or ligand is specified in these columns.
c Five - membered metallocyclic ring.
d Six - membered metallocyclic ring.
e Two crystallographically independent molecules.

Table 2. Crystallographic and structural data for heterobinuclear copper(II) compounds[a].

COMPOUND (colour)	CRYST. CL. SPACE G. Z	a[Å] b[Å] c[Å]	α[°] β[°] γ[°]	CHROMOPHORE	M - L [Å]		Cu - M[Å] Cu - L - M[°]		L - M - L [°]		REF.
[(4-Mepy)Cu(μ-ac)$_2$(μ-CO)$_2$Mo	tr	8.128(2)	106.61(4)	CuO_3C_2N	μacO[b]	2.112(6,5)	Mo	2.558(1)	O,O[b]	94.4(2)	31
(η^5-cp)]	P-1	15.349(8)	98.11(3)		μOC	2.137(9)			O,N	105.5(2,5.0)	
(red)	4	16.086(5)	90.25(3)			2.460(8)			O,Mo	95.7(1,1.6)	
					pyN	1.995(6)			N,Mo	150.4(2)	
				MoC_7O_2	μOC	1.960(9,7)			C,C	76.9(4)	
					cpC	2.325(8,64)			O,O	77.1(2)	
					μacO	2.167(6,2)			O,C	78.5(1,1.2)	
										131.1(3)	
[(H_2O)$_2$Cu(μ-meu)$_2$Pt(NH_3)$_2$]SO_4.	tr	10.398(10)	102.88(6)	CuO_4	H_2O	2.00(2,4)	Pt	2.765(3)	O,O	87.5(8,4.1)	32
4.5H_2O	P-1	10.773(8)	102.62(7)		μmeuO	1.95(2,3)				175.0(8,1.7)	
(blue green)	2	11.772(9)	105.05(7)								
				$Pt^{II}N_4$	H_3N	2.02(2,2)			N,N	90.0(9,2.3)	
					μmeuN	2.00(2,0)				177.1(9,4)	
[(acac)Cu(μ-OSiMe$_3$)$_2$Al	tr	11.107(6)	100.77(5)	$Cu^{I}O_4$	acacO	1.894(9,1)	Al	2.778(4)	O,O	81.3(3)	33a
(OSiMe$_3$)$_2$][d]	P-1	14.956(5)	94.41(5)		μMe$_3$SiO	1.973(8,0)		94.3(4,1)		91.5(4,2)	
(blue violet)	2	19.739(6)	107.08(3)							95.8(4)[c]	
(at 173 K)										172.7(4,5)	
				$Al^{III}O_4$	O	1.711(9,3)			O,O	90.2(4)	
					μO	1.814(9,4)				112.9(4,1.4)	
				CuO_4	acacO	1.891(8,1)	Al	2.784(4)	O,O	80.4((3)	
					μMe$_3$SiO	1.994(8,4)		94.2(4,2)		92.5(4,7)	
										94.5(4)[c]	
										173.3(4)	
				$Al^{III}O_4$	O	1.683(9,5)			O,O	91.1(4)	
					μO	1.802(9,3)				112.6(5,3.1)	

Table 2. Continued 2.

[Cu(μ-salen)Ni(hfac)$_2$]	m	9.287(4)		CuO_2N_2	μO	1.900(2,8)	Ni	2.897(2)	O,O	83.9(1)[d]	33b
(not given)	$P2_1/c$	21.533(2)	94.36(3)		N	1.909(3,5)		93.9(1,2)	N,N	86.3(1)	
(at 123 K)	4	14.733(3)							O,N	94.9(1,7)[c]	
										177.4(1,5)	
				$Ni^{II}O_6$	μO	2.086(2,22)			O,O	75.1(1)	
					hfacO	2.019(2,24)				90.6(1,6)[c]	
										92.3(1,10.7)	
										173.5(1,3.5)	
[(EtOH)$_2$Cu(μ-dbs)Ni]	m	19.946(2)		CuO_4N_2	O	not given	Ni	2.898(2)		not given	33c
(yellow-green)	C2/c	19.340(2)	135.64(1)		N	not given					
	4	18.906(2)			EtHO	2.460(7,0)					
				$Ni^{II}O_2N_2$	O	not given				not given	
					N	not given					
[Cu(μ-(baa)$_2$en)Ni]	m	13.052(2)		CuO_4	O	1.866(5,0)	Ni	2.925(1)	O,O	76.3(2)	34
(deep red)	$P2_1/c$	8.612(2)	98.38(2)		μO	1.919(5,0)		101.9(2,2)		94.5(2,4)[c]	
	4	20.295(4)								168.5(2,2)	
				$Ni^{II}O_2N_2$	μO	1.848(7,21)			O,O	79.8(2)	
					N	1.826(7,1)			N,N	89.3(3)[e]	
									O,N	95.3(3,1)[c]	
										175.3(4,1)	
[(H_2O)Cu(μ-aapen)Ni]	m	12.918(5)		CuO_5	O	1.89(1,0)	Ni	2.931(2)	O,O	78.2(4)	35
(green)	$P2_1/c$	8.481(4)	104.07(4)		μO	1.95(1,10)		100.0(4,8)		91.6(4,6)[c]	
	4	18.782(6)			H_2O	2.27(1)				95.8(4,5.8)	
										164.8(4,1)	
				$Ni^{II}O_2N_2$	μO	1.875(10,5)			O,O	81.9(4)	
					N	1.84(1,0)			N,N	88.1(5)[e]	
									O,N	95.0(5,1)[c]	
										176.5(5,2)	

Table 2. Continued 3.

[(MeOH)Cu(μ-(fsa)$_2$en)VO]	m	11.629(5)		CuO_3N_2	μO	1.912(8,13)	V	2.949(3)	O,O	81.3(3)	36
(red)	$P2_1/n$	13.579(5)	101.2(5)		N	1.883(9,6)		95.1(4,4.3)		95.6(5,1.9)	
	4	12.416(5)			MeHO	2.311(9)			N,N	86.5(5)[c]	
									O,N	95.3(4,2)[c]	
				$V^{IV}O_5$	O	1.574(9,12)			O,O	83.0(4,4.8)	
					O	2.00(1,11)				86.4(4,9.9)	
					μO	1.91(1,11)				106.2(4,6.5)	
[(MeOH)Cu(μ-(fsa)$_2$en)VO]	m	11.636(3)		CuO_3N_2	μO	1.935(9,23)	V	2.989(4)	O,O	80.9(5)	36
(dark blue)	$P2_1/n$	13.612(3)	100.8(4)		N	1.887(9,9)		98.6(7,3)		95.8(8,1.8)	
	4	12.426(3)			MeHO	2.304(8)			N,N	87.6(8)[c]	
									O,N	94.8(6,4)[c]	
				$V^{IV}O_5$	O	1.576(9)			O,O	79.7(5)	
					O	1.897(9,4)				85.4(6,4)	
					μO	1.996(8,18)				102.2(9,5.2)	
[Cu(μ-mesalen)Zn(bhppH$_2$)$_2$].	tr	11.779(6)	62.85(8)	CuO_2N_2	μO	1.901(2,19)	Zn	2.966	O,O	81.9(1)	37
$2CH_2Cl_2$	P-1	13.796(10)	73.51(3)		N	1.942(6,6)		93.8(2,1.0)	N,N	89.5(2)[c]	
(red)	2	17.913(5)	71.57(9)						O,N	93.8(2)[c]	
										175.6(2)	
				$Zn^{II}O_6$	O	2.038(6,33)			O,O	70.6(1)	
					μO	2.154(2,48)				86.7(1,2.0)[c]	
										94.0(1,9.7)	
										168.6(2,7.1)	
[Cu(μ-(fsa)$_2$en)Ni(H$_2$O)$_2$].H_2O	trg	12.8071(4)		CuO_2N_2	μO	1.894(3,2)	Ni	2.9749(6)	O,O	84.2(1)	38
(not given)	$P3_1$				N	1.905(3,3)		98.9(1,0)	N,N	87.5(1)[c]	
(at 153 K)	3	9.8157(8)							O,N	94.1(1,1)[c]	
				$Ni^{II}O_6$	O	1.988(3,2)			O,O	89.1(1,1)[c]	
					μO	2.020(3,2)				77.9(1)	
					H_2O	2.090(3,21)				91.5(1,12.3)	
										170.9(3,8.1)	

Table 2. Continued 4.

[Cu(μ-dat)VO]	m	8.325(20)		CuO_2N_2	μO	1.903(5,1)	V	2.985	O,O	79.7(2)	39
(dark green)	$P2_1/c$	12.117(11)	108.93(6)		N	1.882(6,11)		100.3(2,2)	N,N	90.2(3)[c]	
	4	18.105(8)							O,N	94.8(2,7)[c]	
										172.8(2,4)	
				$V^{IV}O_5$	O	1.600(5)			O,O	86.7(2,0)[c]	
						1.935(5,2)				75.7(2)	
					μO	1.987(5,7)				91.2(2)	
										107.0(2,1.4)	
										144.9(2,1.8)	
[(MeCN)Cu(μ-cypim)Co].$(PF_6)_3$.	m	12.248(2)		CuN_3O_2	MeCN	2.221(11)	Co	2.994(2)	N,N	94.2(4,4.7)	40
EtOH	$P2_1$	11.741(2)	94.91(3)		N	1.980(11,24)		101.9(4,1.5)		108.0(4)[e]	
(dark maroon)	2	17.961(4)			μO	1.974(8,25)			O,O	75.8(3)	
									N,O	89.0(4,9)[c]	
										116.5;140.5(4)	
										163.8(4)	
				$Co^{III}N_4O_2$	N	1.949(10,4)			N,N	86.5(4,4.1)[d]	
						1.977(10)				94.8(4,1.7)	
					μO	1.881(8,17)				177.8(4)	
									O,O	80.3(3)	
									N,O	91.3(4,3.6)	
										94.6(4,1.0)[c]	
										174.2(4,1.3)	
[Cu(S,S-cypim)Mn(EtOH)].	tr	12.374(2)	83.64(3)	CuO_2N_2		not given				not given	40
$(PF_6)_{1.5}.CH_2Cl_2$	P-1	13.453(3)	81.33(3)								
(green-black)	2	16.682(3)	64.66(3)	$Mn^{II}N_4O_3$							

Table 2. Continued 5.

[Cu(μ-(fsa)$_2$en)Mg(H$_2$O)$_2$].H$_2$O (brick red)	tg P3$_1$/21 3	12.84(5) 9.99(5)		CuO$_2$N$_2$	μO N	1.920(3,0) 1.917(3,0)	Mg	3.001(4) 98.2(1)	O,O N,N	85.0(4) 86.0(1)^c 94.5(1)^c 179.1(1)	41
				Mg^{II}O$_6$	O μO H$_2$O	1.978(2,0) 2.049(3,0) 2.115(3,0)			O,O	78.2(1) 85.4(1)^c 110.8(1) 163.4(1)	
[Cu(μ-(prp)$_2$en)Fe(hfac)$_2$] (not given)	m P2$_1$/c 4	13.576(7) 20.365(5) 12.606(7)	97.70(5)	CuO$_2$N$_2$	μO N	1.88(1,10) 1.94(1,3)	Fe	3.00 96.1(5,9)	O,O N,N	83.0(5) 89.5(6)^c 93.7(5,1.6)^c 174.8(6,3.3	42
				Fe^{II}O$_6$	O μO	2.07(1,4) 2.145(10,35)			O,O	70.9(4) 85.5(5,3)^c 93.7(5,8.5) 169.1(5,4.7)	
[Cu(μ-(fsa)$_2$en)Co(H$_2$O)$_2$].H$_2$O (red purple)	trg P3$_2$21 3	12.845(2) 9.962(1)		CuO$_2$N$_2$	μO N	1.909(5,0) 1.915(6,0)	Co	3.025(2) 99.1(2)	O,O N,N O,N	84.7(2) 85.9(3)^c 94.7(2)^c	43
				Co^{II}O$_6$	O μO H$_2$O	1.995(5,0) 2.065(4,0) 2.137(5,0)			O,O	77.1(2) 86.2(2)^c 94.2(2,16.5)	
[Cu(μ-(prp)$_2$pr)Co(hfac)$_2$] (not given)	tr P-1 2	11.821(3) 12.402(5) 13.885(6)	91.64(3) 95.64(3) 116.75(3)	CuO$_2$N$_2$	μO N	1.91(1,4) 2.015(20,25)	Co	3.03 99.9(5,6)	O,O N,N O,N	78.2(5) 94.8(7)^c 93.8(6,1.6)^c 169.5(7,2.2)	42
				Co^{II}O$_6$	O μO	2.055(10.45) 2.04(1,5)			O,O	72.2(4) 87.7(5,1)^c 93.5(5,8.4) 168.3(5,6.3)	

Table 2. Continued 6.

[(H$_2$O)Cu(μ-(fsa)$_2$en)Co(H$_2$O)].	or	20.549(5)		CuO$_2$N$_2$	μO	1.888(4,6)	Co	not given	O,O	83.3(3)	44
4H$_2$O	Pbca	15.948(4)			N	1.885(5,3)		100.0(4,2)		89.4(3,1.4)	
(blue-black)	8	13.307(5)			H$_2$O	2.784(5)			N,N	87.6(4)[c]	
									O,N	94.5(4,7)[c]	
										88.4(4,5.9)	
										176.8(4,1)	
				Co^{II}O$_5$	O	1.948(4,7)			O,O	89.1(3)[c]	
					μO	2.000(4,22)				75.6(3,3.2)	
					H$_2$O	2.071(5)				94.4(3,9.8)	
										161.2(3,3.0)	
[Cu(μ-salen)Mg(hfac)$_2$].2H$_2$O	tr	9.501(7)	83.7(1)	CuO$_2$N$_2$	μO	1.915	Mg	3.031(5)		not given	45
(not given)	P-1	15.43(1)	112.8(1)		N	not given		not given			
	2	12.042(8)	107.8(1)								
				Mg^{II}O$_6$	μO	2.053			O,O	76.2-97.6(2)	
						2.109(5)				164.2-174.5(2)	
[Cu(μ-dp)Ni(H$_2$O)(dmf)].(ClO$_4$)$_2$.	or	15.62(1)		CuO$_2$N$_2$	μO	1.899(7,10)	Ni	3.038(2)	O,O	83.8(3)	46
dmf	P2$_1$2$_1$2$_1$	16.49(1)			N	1.897(9,6)		99.9(3,1)	N,N	84.6(5)[c]	
(brown)	4	14.75(1)							O,N	95.6(4,1.4)	
				Ni^{II}O$_4$N$_2$	μO	2.068(7,6)			O,O	75.7(3)	
					N	2.051(9,5)				90.9(3,4.9)	
					H$_2$O	2.101(7)			N,N	107.2(4)[f]	
					dmfO	2.077(8)			O,N	88.8(3,2.6)	
[Cu(μ-salen)Co(hfac)$_2$]	tr	18.412(5)	113.90(2)	CuO$_2$N$_2$	μO	1.893(9,4)	Co	3.060(1)	O,O	83.9(3)	47
(dark red)	P-1	9.419(3)	123.25(2)		N	1.895(13,11)		99.8(2,6)	N,N	84.7(4)[c]	
	2	12.044(3)	89.20(2)						O,N	95.9(3,1.3)[c]	
										174.9(3,1.9)	
				Co^{II}O$_6$	O	2.066(10,29)			O,O	86.6(3,1.0)[c]	
					μO	2.105(10,15)				73.9(3)	
										92.8(3,8.0)	

Table 2. Continued 7.

[Cu(μ-$C_2H_4O_2$)$_2$Ba($C_2H_6O_2$)$_3$] (deep blue) (at 203 K)	m $P2_1/n$ 4	12.127(4) 11.913(1) 12.540(4)	 102.47(1)	CuO_4	μO O	1.927(3,5) 1.915(3,5)	Ba	not given	O,O	88.5(1,3)[c] 91.6(1,1.2)	10
				$Ba^{II}O_6$	μO O	2.743(3,11) 2.803(3,60)			O,O	not given	
[Cu(μ-mch)(μ-ac)Mn]BPh_4 (blue-green)	m $P2_1/n$ 4	17.508(8) 15.539(5) 16.515(5)	 106.30(3)	CuO_3N_2	μmchO N μacO	1.934(8,6) 1.940(9,4) 2.178(7)	Mn	3.122(2) 96.2(3,6)	O,O N,N O,N	85.3(3) 94.8(3,3.2) 84.2(4)[c] 94.1(3,9)[c] 97.3(3,1)	48a
				$Mn^{II}O_3N_3$	μmchO N μacO	2.250(7,24) 2.23(1,0) 2.070(70			O,O N,N O,N	71.3(2) 91.2(3,1.2) 74.6(3,3)[c] 107.8(3) 77.3(3,7)[c] 84.0(3) 119.6(3,4.4)	
[Cu(μ-salen)Mn(hfac)$_2$] (red)	tr P-1 2	15.584(4) 12.039(3) 9.470(2)	113.83(2) 107.17(3) 84.28(30	CuO_2N_2	μO N	1.887(8,8) 1.921(11,6)	Mn	3.123 100.2(3,6)	O,O N,N O,N	85.4(3) 84.8(5)[c] 95.1(4,1.1)[c] 174.6(4,2.1)	48b
				$Mn^{II}O_6$	μO hfacO	2.176(8,11) 2.157(10,19)			O,O	72.0(3) 82.3(3,1.4)[c] 94.2(3,12.3) 165.9(3,6.5)	

Table 2. Continued 8.

[(MeOH)Cu(μ-(fsa)$_2$en)FeCl	m	10.619(60		CuO_3N_2	μO	1.929(6,3)	Fe	not given	O,O	82.3(3)	49
(H_2O)].MeOH	$P2_1/c$	11.638(3)	101.59(3)		N	1.911(8,4)		100.0(3,7)		95.8(3,1.6)	
(dark red)	4	18.428(9)			MeHO	2.277(7)			N,N	85.8(4)[c]	
									O,N	94.5(3,5)[c]	
										96.8(3,1.0)	
										167.2(3,5)	
				$Fe^{III}O_5Cl$	O	1.930(7,8)			O,O	77.5(2)	
					μO	2.028(6,22)				88.8(3,1.5)[c]	
					H_2O	2.150(7)				85.3(3,1.9)	
					Cl	2.303(3)				107.2(3)	
										162.9(3,4)	
[(SCN)Cu(μ-dm) Zn(NCS)].dmf	m	14.990(6)		CuN_3O_2	N	1.936(10,12)	Zn	3.166(2)	N,N	84.0(4)[c]	50
(deep green)	$P2_1/n$	18.640(4)	102.55(2)		μO	1.932(7,11)		102.4(3,1.3)		94.9(4,8)	
	4	12.016(2)			SCN	2.202(40			O,O	82.1(3)	
									N,O	94.6(3,1)[c]	
										102.0(4,5.4)	
				$Zn^{II}N_3O_2$	N	2.110(9,8)			N,N	103.5(4,7.5)	
					μO	2.130(7,25)			O,O	73.1(2)	
					SCN	2.003(10)			O,N	83.6(3,3)[c]	
										100.6(3,2.7)	
[Cu(μ-prp)$_2$pr)Fe(hfac)$_2$]	tr	11.630(4)	90.83(5)	CuO_2N_2	μO	1.913(12,6)	Fe	3.20	O,O	76.6(6)	42
(not given)	P-1	12.443(6)	95.26(4)		N	2.005(12,72)		100.0(4,2)	N,N	97.2(4)[c]	
	2	13.897(4)	116.19(4)						O,N	93.7(6,3.7)[c]	
										166.6(6,3.0)	
				$Fe^{II}O_6$	μO	2.04(1,5)			O,O	69.3(4)	
					O	2.027(12,36)				85.5(5,1)[c]	
										93.6(5,10.0)	
										169.1(4,6.3)	

Table 2. Continued 9.

[Cu(μ-bdmap)$_2$(μ-F$_3$ac).Ca(F$_3$ac)	m	13.234(4)		CuO$_3$N$_2$	μO	1.909(6,17)	Ca	3.262(2)	O,O	83.1(3)	51
H$_2$O]	P2$_1$/n	15.401(3)	94.79(2)		N	2.081(9,3)		99.5(3,1.6)	N,N	109.8(4)	
(dark blue)	4	14.445(2)			μF$_3$acO	2.397(8)			O,N	83.0(3,1)[c]	
										164.1(3,1.5)	
				Ca^{II}O$_5$N$_2$	μO	2.350(7,30)			O,O	72.4(2,7.2)	
					N	2.668(9,27)				92.8(3,3.1)	
					μF$_3$acO	2.370(8)				147.0(2,7)	
					F$_3$acO	2.415(7)				173.9(3)	
					H$_2$O	2.386(7)			N,N	161.8(3)	
									O,N	65.7(3,4)[d]	
										88.7(3,7.8)	
										130.7(3,5)	
[Cu(μ-fms)Pb(ClO$_4$)(dmf)].ClO$_4$	tr	12.531(5)	106.65(3)	CuO$_2$N$_2$	μO	1.89(1,1)	Pb	3.344(3)	O,O	84.3(5)	52
(green)	P-1	13.961(4)	100.07(3)		n	1.91(2,3)		95.7	N,N	87.7(7)[c]	
	2	10.697(3)	102.38(3)						O,N	94.0(6,9)[c]	
										177.0(8,1.7)	
				Pb^{II}O$_4$N$_2$	μO	2.53(1,4)			O,O	60.1(4)	
					n	2.48(2,3)				83.6(5,4.0)	
					O$_3$ClO	2.61(2)				139.0(6,6.4)	
					dmfO	2.60(1)			N,N	79.9(5)[c]	
									O,N	65.7(6,3)[c]	
										70.7(6,4.8)[c]	
										105.8(5,2.9)	
										143.1(6)	

Table 2. Continued 10.

[(F_3ac)Cu(μ-bdmap)Pr(hfac)	m	18.205(5)		CuO_3N_2	μO	1.94(1,1)	Pr 3.355(3)	O,O	81.4(6)	53
(mtfd)]	Cc	18.350(6)	94.96(2)		N	2.10(1,1)	101.2(5,3)		103.3(7,1.8)	
(green blue)	4	15.924(4)			F_3acO	2.21(2)		N,N	106.0(8)	
								O,N	84.4(6,3)[e]	
									160.7(7,8)	
				$Pr^{III}O_8$	O	2.40(1,0)		O,O	71.8(5,3.5)	
					hfacO	2.46(2,5)			84.2(5,8.3)	
					mtfdO	2.58(2,3)			121.0(5,3.8)	
									141.0(5,7.4)	
[(F_3ac)Cu(μ-bdmap)Y(hfacac)	m	18.218(4)		CuO_3N_2		not given	not given		not given	53
(mtfd)]	Cc	18.316(7)	94.86(2)							
(not given)	4	15.706(3)		$Y^{II}O_8$						
[Cu(μ-bpmp)(μ-pr)$_2$Fe].(PF_6)$_2$	or	20.849(6)		CuO_3N_3	μbpmpO	2.01(1)	Fe 3.401(4)		not given	54
(not given)	Pnma	17.411(4)			N	2.06(1,3)	119.3(5)			
(at 143 K)	4	12.587(3)				2.19(1)				
					μprO	1.94(1)				
						2.31(1)				
				$Fe^{III}O_3N_3$	μbpmpO	1.93(1)			not given	
					N	2.14(1,2)				
					μprO	1.90(2)				
						2.11(1)				
[ClCu(μ-dppn)(μ-NO)Ir(PPh_3)$_2$Cl]	m	35.777(17)		CuN_3Cl	μON	2.01(2)	Ir 3.416(5)	N,N	80(1)[e]	55
(PF_6)$_2$	$P2_1/a$	12.393(9)	100.65(7)		μdppnN	1.99(2,1)	117(1)			
(blue)	4	13.481(8)			Cl	2.15(1)				
				$Ir^{III}N_3P_2Cl$	μON	1.98(2)		N,N	78(1)[e]	
					μdppnN	2.02(2,1)				
					Ph_3P	2.41(1,2)				
					Cl	2.33(1)				

Table 2. Continued 11.

Compound (colour)	Cryst. cl. / Space group / Z	a / b / c [Å]	β [°]	Chromophore	M–L [Å]	M–M [Å] / M–L–M [°]	L–M–L [°]		Ref.
$[Cu(\mu\text{-bmm})Gd(NO_3)_3].Me_2CO$ (not given)	m $P2_1/c$ 4	9.799(1) 18.926(3) 16.012(1)	96.530(8)	CuO_2N_2	μO 1.892(5,13) N 1.930(7,12)	Gd 3.428(1) O 106.7(2,8)		not given	56
				$Gd^{III}O_{10}$	μbmmO 2.368(5,31) bmmO 2.620(5,6) O_2NO 2.491(6,39)			not given	
$[(O_3ClO)Cu(\mu\text{-nup})Ni(Me_3[12]N_3)]ClO_4.H_2O$[d] (red)	or $Pna2_1$ 8	20.046(5) 10.462(2) 35.397(7)		CuN_3O_2	N 1.97(2,2) O 1.89(2) O_3ClO 2.79(3)	Ni 3.440(4) 120.1(8)	N,N O,O N,O	79.8(8)[c] 96.9(8)[c] 175.3(8) 80.0(7) 95.3(7)[c] 92.2(7,9.7) 167.8(7)	57a
				$Ni^{II}N_3O_2$	N 2.02(2,4) O 2.05(2,3)		N,N O,O N,O	96.1(8,5.6)[c] 83.9(6) 93.0(7,5.3) 165.7(7,6.0)	
				CuN_3O_2	N 1.95(2,3) O 1.92(2) O_3ClO 2.56(2)	Ni 3.452(4) 119.3(7)	N,N O,O N,O	82.1(8)[c] 97.0(8)[c] 175.0(9) 82.7(7) 94.1(8)[c] 90.0(8,7.8) 168.9(7)	
				$Ni^{II}N_3O_2$	N 2.03(2,6) O 2.06(2,3)		N,N O,O N,O	96.9(9,6.2)[c] 85.5(6) 92.3(9,6.1) 164.5(8,6.2)	

Table 2. Continued 12.

[(H$_2$O)Cu(μ-ala)$_4$Y(H$_2$O)$_4$].	m	14.370(4)		CuO$_5$	O	1.937(4,4)	Y	3.491	O,O	91.4(1,4.5)	57b
(ClO$_4$)$_5$.3H$_2$O	P2$_1$	8.920(1)	91.36(2)		H$_2$O	2.272(3)				174.1(1,9)	
(blue)	2	16.651(2)									
				Y^{III}O$_8$	O	2.292(3)			O,O	75.5(1,8.0)	
						2.319(3)				103.1(1)	
					H$_2$O	2.409(3,3)				119.0(1)	
										142.0(1,6.0)	
[(dmso)Cu(μ-aapen)UO$_2$(dmso)].	m	26.22(2)		CuO$_3$N$_2$	μO	1.94(1,2)	U	3.51	O,O	84.4(6)	57c
dmso	P2$_1$/a	14.79(2)	104.65(5)		N	1.93(1,1)		103(1,1)		95(1,1)	
(brown-red)	4	8.10(1)			dmsoO	2.32(1)			N,N	85.5(7)[c]	
									O,N	94.0(7,5)[c]	
										96.0(1,4)	
										168.0(1,4)	
				U^{VI}O$_7$	O	1.81(1,1)			O,O	62.9(4)	
					aapenO	2.24(1,1)				70.2(5,1.2)	
					μaapenO	2.51(1,2)				78.4(6,1)	
					dmsoO	2.39(1)				178.0(7)	
[(bdmmpH)Cu(μ-bdmmp)(μ-OH)	m	14.304(8)		CuO$_3$N$_2$	μHO	2.033(8)	Pr	3.538(2)	O,O	86.5(3,5.5)	58
Pr(hfac)$_3$]	P2$_1$/n	19.122(6)	100.21(3)		μmpO	1.958(8)	μHO	103.9(3)		172.1(4)	
(dark green)	4	20.059(4)			mpN	2.10(1)	μO	107.2(3)	N,N	117.4(4)	
					mpHO	1.915(8)			O,N	92.5(4,1.0)[c]	
					mpHN	2.30(1)				91.5(4,2.5)	
										106.1;136.5(4)	
				Pr^{III}O$_8$N	μHO	2.450(8)			O,O	64.3	
					μmpO	2.424(8)				153.9(3)	
					mpN	2.90(1)					
					hfacO	2.48(1,2)					

Table 2. Continued 13.

Compound				Chromophore	L	M–L	M		L,L	L–M–L	Ref.
[Cu(μ-dopn)(μ-ac)Mn(tmac)]	or	12.275(4)		CuN_4O	N	1.958(7)	Mn	3.539	N,N	80.0(3,1.0)[e]	59
$(ClO_4)_2.2H_2O$	$P2_12_12_1$	14.171(9)				1.991(7,5)				96.2(2)[e]	
(deep brown)	4	19.780(3)			μacO	2.166(5)				98.2(2)	
										162.7(3,2)	
									N,O	98.6(2,3.5)	
				$Mn^{III}O_3N_3$	μacO	2.093(6)			O,O	96.8(2,9)	
					O	1.909(5,7)			N,N	81.3(2,1.2)[e]	
					N	2.116(6,12)			O,N	90.4(2,2.9)	
						2.266(6)				169.0(2,2.0)	
[$(ClO_4)(Me_5$dien)Cu(μ-OH)Fe	tr	12.08(1)	106.17(2)	CuN_3O_2	dienN	2.030(4,9)	Fe	3.570	O,N	91.1(1,1.7)	60
(oep)](ClO_4)	P-1	14.20(1)	97.58(2)		μHO	1.954(4)		175.3		98.9(2,6.8)	
(dark red)	2	16.30(1)	109.21(2)		O_3ClO	2.426(4)			N,N	157.0(2)	
				$Fe^{III}N_4O$	oepN	2.044(4,7)			O,N	94.7(2)	
					μHO	1.929(4)				108.2(2)	
[(pnaoH)Cu(μ-O)ReO_3]	or	11.989(6)		CuN_4O	N	1.975(10,5)	Re	not given	N,N	81.4(5)[e]	61
(purple)	Pmcn	12.430(6)			μO	2.40(1)				96.3(6)[e]	
	4	13.258(7)								97.4(8)	
									N,O	97.0(6,2.0)	
				$Re^{VII}O_4$	O	1.68(1,4)			O,O	109.5(7,1.6)	
					μO	1.72(1)					
[(terpy)$(N_3)Cu_{1.5}$(μ-1,1-$N_3)_2Ni_{0.5}$	m	10.195(5)		$M^{II}N_6$	terpyN	2.00(3,5)		3.59(1)	N,N	79.4(1,2)[e]	62
(N_3)(terpy)]$.2H_2O$	$P2_1/c$	9.996(3)	91.10(5)		N_3	2.26(2)		95.1(1)		82.3(1,4.8).	
(dark green)	4	15.866(8)			μN	2.01(3,4)				95.8(1,7.7)	
										160.3(1,1.7)	
[(tpma)Cu(μ-O)Fe(F_8tpp)](ClO_4).	or	20.017(7)		CuN_4O	N	1.996(8,12)	Fe	3.596(2)	N,N	158.4(3)	63
$3C_7H_8.2MeCN$	$P2_12_12_1$	24.530(8)				2.136(7,36)		178.2(4)	N,O	176.7(3)	
(red purple)	4	15.451(4)			μO	1.856(5)					
				$Fe^{III}N_4O$	N	2.103(7,4)			N,N	149.5(2,1.3)	
					μO	1.740(5)					

Table 2. Continued 14.

$[Cl_3Cu(\mu\text{-}Cl)Mg(15\text{-}crown\text{-}5)$	m	9.249(3)		$CuCl_4$	Cl	not given				not given	15
$(MeCN)]$	$P2_1/n$	16.931(7)			μCl	not given					
(yellow)	4	13.198(4)	94.61(2)								
				$Mg^{II}O_5NCl$	O	2.166(4,49)			O,O	72.1(1,4)[c]	
					MeCN	2.172(5)			N,O	88.5(2,4.9)	
					μCl	2.494(2)			Cl,O	91.6(1,6.2)	
									Cl,N	174.1(1)	
$[(ae)Cu(\mu\text{-}im)Ni(ae)]ClO_4$	tr	10.227(2)	103.34(1)	CuN_3O	μimN	1.950(4)	Ni	3.632(1)	N,N	84.4(2)[c]	64
(red)	P-1	13.218(2)	100.27(1)		aeN	1.939(5,28)				94.3(2)	
	2	8.271(1)	76.87(1)		aeO	1.874(2)				174.9(2)	
									N,O	87.8(2)	
										93.9(2)[c]	
										175.0(2)	
				$Ni^{II}N_3O$	μimN	1.934(1)			N,N	84.0(2)[c]	
					aeN	1.920(5,28)				94.5(2)	
					aeO	1.855(1)				178.0(2)	
									N,O	87.5(2)	
										94.0(2)[c]	
										176.5(2)	
$[Cu(\mu\text{-}bpmp)(\mu\text{-}ac)Fe(MeO)]$	tr	13.135(3)	93.62(2)	CuN_3O_2	bpmpN	2.050(6,30)	Fe	3.641(1)		not given	54
$(BPh_4)_2.0.22acH$	P-1	13.503(3)	93.81(2)		μbpmpO	2.176(4)		122.5(2)			
(red green)	2	20.940(5)	92.76(2)		μacO	1.945(4)					
(at 146 K)											
				$Fe^{III}O_3N_3$	MeO	2.016(4)				not given	
					μacO	1.849(4)					
					μbpmpO	1.975(4)					
					bpmpN	2.154(5,22)					
						2.255(5)					

Table 2. Continued 15.

$[Cu(\mu\text{-eob})Ba(CF_3SO_3)_2]$[d]	m	12.213(6)		CuO_2N_2	O	1.94(-,4)	Ba	3.69		not given	65
(dark brown)	Pn	10.558(2)	102.83(1)		N	1.95(-,1)					
	4	26.831(9)									
				$Ba^{II}O_9$	O	2.65-2.85				not given	
					CF_3O_2SO	2.73-3.03					
				CuO_2N_2	O	1.91(-,1)				not given	
					N	1.97(-,1)					
				$Ba^{II}O_9$	O	2.66-2.85				not given	
					CF_3O_2SO	2.56-2.94					
$[Cu(\mu\text{-hot})Ba(CF_3SO_3)(H_2O)].$	tr	12.915(1)	61.87(1)	CuO_2N_2	O	1.94(-,1)	Ba	3.73		not given	65
$CF_3SO_3.H_2O$	P-1	12.501(3)	81.73(1)		N	1.97(-,3)					
(green)	2	12.020(1)	78.47(1)								
				$Ba^{II}O_9$	O	2.68-2.91				not given	
					CF_3O_2SO	2.81(-,5)					
					H_2O	2.82					
$[(en)_2(H_2O)Cu(\mu\text{-edta})Ni].4H_2O$	m	10.015(1)		CuN_4O_2	enN	1.984(5,8)	Ni	3.844	N,N	85.4(2,1.8)[e]	66
(not given)	$P2_1$	11.816(2)	104.43(1)			2.048(5,2)		118.4(2)		94.3(2,9)	
	2	10.386(1)			μO	2.388(4)				175.3(2,2.3)	
					H_2O	2.632(5)			O,O	173.3(2)	
									N,O	90.0(2,7.0)	
				$Ni^{II}O_4N_2$	O	1.991(5)			O,O	93.9(2,18.6)	
						2.084(5,6).			O,N	82.3(2,2.7)[e]	
					μO	2.084(4)				97.6(2,7)	
					N	2.060(5,6)				164.0(2,2.4)	

Table 2. Continued 16.

$[Cu(\mu\text{-}cdta)Zn(H_2O)_5].H_2O$	or	10.868(3)		CuO_4N_2	O	1.991(5,17)	Zn	not given	O,O	93.6(2,13.2)	67
(blue)	$Pbc2_1$	11.139(8)				2.254(6,12)				178.9(2)	
	4	17.432(5)			N	2.037(5,23)			N,N	88.8(2)[e]	
									O,N	82.7(2,2.0)[e]	
										97.8(2,2.1)	
										165.4(2,1.8)	
				$Zn^{II}O_6$	O	2.048(5)			O,O	90.0(2,7.5)	
					H_2O	2.059(5,2)				173.4(2,1.2)	
						2.148(5,56)					
$[(H_2O)_4Cu(\mu\text{-}cdta)Ni].3H_2O$	tr	8.685(4)	96.43(4)	CuO_6	μO	1.964(3)	Ni	not given	O,O	51.2(1)[g]	68
(blue)	P-1	11.415(6)	96.60(3)		O	2.855(4)				90.7(2,13.1)	
	2	12.841(4)	102.23(6)		H_2O	1.948(4,17)				120.3;146.2(2)	
						2.257(5)				171.4(2)	
				$Ni^{II}O_4N_2$	O	2.045(4,17)			O,O	88.3(2,6.7)	
					μO	2.097(3)				114.1(2)	
					N	2.082(4,22)			N,N	86.6(2)[e]	
									O,N	82.7(2,2.9)[e]	
										100.6(2,1.4)	
										161.2(2,2.2)	
$[(H_2O)Cu(\mu\text{-}dopn)Cr(MeO)$	m	13.096(3)		CuN_4O	N	1.933(41,57)	Cr	3.86	N,N	81.6(16,4)[e]	59
$(tmac)](ClO_4)_2.H_2O$	$P2_1/n$	17.933(4)	113.49(3)		H_2O	2.285(19)				97.5(16)[e]	
(red-brown)	4	15.994(3)								97.1(15)	
										167.5(13,4.6)	
									N,O	96.3(11,3.7)	
				$Cr^{III}O_3N_3$	dopnO	1.944(27,24)			O,O	97.5(10,3.4)	
					MeO	1.895(22)			N,N	81.4(12,1.0)[e]	
					N	2.168(35,5)			O,N	90.0(12,3.1)	
										168.6(13,1.9)	

Table 2. Continued 17.

[(dced)Cu(μ-dced)Fe(tpp)(thf)].	tr	11.622(2)	83.60(1)	CuS_4	S	not given					69a
[Fe(tpp)(thf)$_2$].2thf	P-1	13.592(2)	74.90(1)								
(dark)	1	16.696(2)	88.60(1)	$Fe^{III}N_4OS$		not given					
(at 143 K)	tr	11.463(3)	82.37(2)	CuS_4	S	2.250(3,10)	Fe	3.924(4)	S,S	89.9(1,2.2)[c]	
	P-1	13.351(3)	74.07(2)		μS	2.307(3)		112.5(1)		175.1(1,1.3)	
	1	16.643(4)	89.28(2)								
				$Fe^{III}N_4OS$	tppN	1.997(9,13)			N,N	90.0(3,1.1)[c]	
					thfO	2.212(7)				175.3(3,1)	
					μS	2.411(3)			N,O	88.6(3,3.2)	
									N,S	92.2(2,5.0)	
									O,S	172.3(2)	
[(mnt)Cu(μ-mnt)Fe(tpp)(thf)][Ga	tr	11.478(4)	82.08(2)	CuS_4	S	2.253(3,34)	Fe	3.938(4)	S,S	90.0(1,2.4)[c]	69a
(tpp)(thf)$_2$].2thf	P-1	13.375(4)	73.93(2)		μS	2.308(3)		112.3(1)		175.6(1,1.6)	
(dark)	1	16.665(5)	89.19(2)								
(at 143 K)				$Fe^{III}N_4OS$	tppN	2.004(8,27)			N,N	90.0(3,1.1)[c]	
					thfO	2.219(7)				175.6(2,2)	
					μS	2.434(3)			N,O	88.5(3,2.8)	
									N,S	92.2(2,4.8)	
									O,S	173.0(2)	
[(mnt)Cu(μ-mnt)Fe(tpp)(thf)][Al	tr	11.511(2)	84.19(1)	CuS_4	S	2.255(4,15)	Fe	3.989(4)	S,S	90.1(1,2.3)[c]	69a
(tpp)(thf)(H_2O)].3thf	P-1	13.489(3)	75.22(1)		μS	2.283(3)		115.1(1)		174.1(2,1)	
(dark)	1	16.481(3)	89.09(2)								
(at 203 K)				$Fe^{III}N_4OS$	tppN	2.006(8,12)			N,N	90.0(3,1.2)[c]	
					thfO	1.998(8)				174.5(3,5)	
					μS	2.444(3)			N,O	88.9(3,1.1)	
									N,S	92.7(2,4.2)	
									O,S	173.2(2)	

Table 2. Continued 18.

[(bnpy$_2$)(OClO$_3$)Cu(μ-F)Fe	tr	12.673(3)	93.69(2)	CuN_3FO	N	1.979(4,7)	Fe	3.956(1)	N,N	94.9(2,1.2)[c]	69b
(oep)](PF$_6$)	P-1	13.832(4)	105.64(2)			2.041(4)		171.9(1)		169.7(2)	
(black)	2	16.901(5)	92.91(2)		μF	2.101(3)			F,N	90.0(1,8)	
					O$_3$ClO	2.139(3)				106.5(1)	
									O,N	84.9(2,8)	
										156.8(1)	
									O,F	96.7(1)	
				$Fe^{III}N_4F$	N	2.049(4,8)			N,N	88.2(1,5)[c]	
					μF	1.865(3)				159.4(1,5)	
									F,N	100.4(1,2.3)	
[Cu(μ-acacP)$_2$PtCl$_2$].2CH$_2$Cl$_2$.	tr	11.914(6)	102.34(4)	CuO_4	O	1.890(6,11)	Pt	3.97	O,O	90.7(3,4.3)	70
.H$_2$O	P-1	11.646(6)	103.71(4)							165.5(3,1)	
(green)	2	21.125(10)	99.83(4)								
				$Pt^{II}Cl_2P_2$	Cl	2.361(2,11)			Cl,Cl	86.29(8)	
					P	2.283(3,9)			P,P	104.98(7)	
									Cl,P	84.4(1,1.3)	
										168.7(1,6)	
[(mnt)Cu(μ-mnt)Fe(tpp)(thf)][Fe	tr	11.622(2)	83.60(1)	CuS_4	S	2.248(5,41)	Fe	4.039(1)		not given	71
(tpp)(thf)$_2$].2thf	P1	13.582(2)	74.91(1)		μS	2.382(2)		113.4(1)			
(dark)	1	16.696(2)	88.60(1)								
				$Fe^{III}N_4OS$	tppN	2.01(1,2)				not given	
					thfO	2.38(1)					
					μS	2.450(1)					
[(mnt)Cu(μ-mnt)Ni(tba)(MeCN)].	or	16.691(2)		CuS_4	S	2.251(4,23)	Ni	not given	S,S	91.1(1,8)[c]	72a
.MeCN	Pbca	32.256(3)									
(black)	8	14.581(1)		$Ni^{II}N_6$	mntN	2.141(7)			N,N	90.7(3,1)[c]	
					tbaN	2.030(7,10)				90.0(3,6.9)	
					MeCN	2.193(7)					
[Cu(μ-dpb)MnCl].C$_8$H$_{10}$	tr	13.232(5)	103.22(3)	CuN_4	N	1.98(1,1)	Mn	4.127(3)	N,N	90.0(6,2.2)[c]	72b
(not given)	P-1	15.228(6)	101.63(3)			2.015(10,15)					
	2	18.997(5)	94.86(3)								
				$Mn^{II}N_4Cl$	N	2.02(1,2)			N,N	90.0(4,2.8)	
					Cl	2.382(5)			N,Cl	97.2(4,2.7)	

Table 2. Continued 19.

[(bnpy$_2$)(MeCN)Cu(μ-F)Fe	tr	13.96(1)	106.0(1)	CuN$_4$F	bnpy$_2$N	2.00(1,1)	Fe	4.047(3)	N,N	87.7(6,2.0)	69b
(oep)](ClO$_4$)(PF$_6$).MeCN	P-1	14.08(2)	100.54(8)			2.08(1)		166.2(5)		92.2(6,4.6)^c	
(dark brown)	2	18.49(2)	111.74(8)		MeCN	2.01(2)				172.7(6,1.9)	
					μF	2.22(1)			F,N	92.3(5,5.6)	
				Fe^{III}N$_4$F	N	2.03(1,1)			N,N	88.0(6,6)^c	
						2.06(1,1)				158.6(5,9)	
					μF	1.86(1)			F,N	100.8(5,6)	
[(Me$_5$dien)Cu(μ-ac)Fe(oep)]	tr	13.777(3)	96.82(3)	CuN$_3$O	dienN	2.01(1.4)	Fe	4.430	N,N	86.2(5,6)^c	73
(ClO$_4$)$_2$.CH$_2$Cl$_2$	P-1	13.847(3)	108.06(3)		μacO	1.936(8)				165.1(4)	
(red)	2	17.608(4)	114.32(3)						O,N	93.7(4,1.9)	
(at 223 K)										177.8(4)	
				Fe^{III}N$_4$O	oepN	2.040(8,16)			N,N	88.1(4)^c	
					μacO	1.961(7)			O,N	100.0(3,1.7)	
[(en)$_2$Cu(μ-Cl)MnCl$_3$]	m	11.276(3)		CuN$_4$Cl	enN	2.012(5,22)	Mn	4.612(2)	N,N	84.6(2,2)^c	74a
(blue violet)	P2$_1$/a	13.904(3)	91.26(2)		μCl	2.804(2)		125.3(1)		95.3(2,1.2)	
	4	9.055(3)								175.4(2,2.1)	
									N,Cl	92.2(1,3.1)	
				Mn^{II}Cl$_4$	Cl	2.367(2,24)			Cl,Cl	109.5(1,4.2)	
					μCl	2.385(2)					
[(Me$_6$tren)Cu(μ-NC)Fe(oep)(py)]	tr	13.833(3)	94.90(2)	CuN$_5$	trenN	2.10(1,11)	Fe	4.94	N,N	86.4(6,1.4)^c	74b
(SbF$_6$)$_2$.Me$_2$CO	P-1	16.552(4)	94.66(2)		μCN	1.88(1)				93.6(6,5)	
(dark violet)	4	31.532(6)	112.52(2)							119.6(5,1.3)	
										178.0(6)	
				Fe^{III}N$_5$C	oepN	2.00(1,2)			N,N	90.2(4,9)^c	
					pyN	2.05(1)			N,C	89(1)	
					μCN	1.92(2)				178.0(5)	

Table 2. Continued 20.

[(Me$_5$dien)(SO$_3$CF$_3$)Cu(μ-NC)Fe	tr	11.941(2)	84.08(2)	CuN$_4$O	μCN	1.949(5)	Fe	4.98	N,N	86.5(2,1)[c]	74b
(oep)(py)](SO$_3$CF$_3$)	P-1	13.576(3)	80.38(2)		dienN	2.041(4,22)				92.0(2,7)	
(dark violet)	2	19.280(5)	73.60(2)		CF$_3$O$_2$SO	2.239(6)				156.3(2)	
										172.5(2)	
									N,N	95.4(2,4.0)	
				Fe^{III}N$_5$C	oepN	1.989(5,5)			N,N	90.2(2,1.0)[c]	
					pyN	2.066(4)			N,C	89(2)	
					μNC	1.898(5)				178.1(2)	
[(O$_3$ClO)Cu(μ-nup)Ni(cth)].	m	18.046(3)		CuN$_4$O	N	1.973(7,10)	Ni	4.972(1)	N,N	80.9(2,8)[c]	57a
ClO$_4$.H$_2$O	P2$_1$/c	12.541(1)	101.08(1)		O$_3$ClO	2.67(2)				97.6(2)[c]	
(brown red)	4	19.471(4)								110.0(7)	
										172.3(7,4.0)	
									N,O	94.8(7,9.5)	
										166.6(2)	
				Ni^{II}N$_4$O$_2$	O	2.08(2)			N,N	84.4(2,1)[c]	
					cthO	2.09(1)				94.6(2,7.4)	
					cthN	2.077(5)				165.6(2,5.9)	
						2.152(6,16)			O,O	97.2(6)	
									N,O	73.4(2,3)[c]	
										91.7(6,6.2)	
										172.4(6,2.4)	
[Cu(μ-Cl$_4$obbz)Mn(H$_2$O)$_5$]	m	10.351(4)		CuO$_2$N$_2$	O	1.921(5,16)	Mn	5.195(1)	O,O	86.8(2)	75
(dark violet)	P2$_1$/n	19.130(4)	97.82(2)		N	1.917(5,3)			N,N	85.2(2)[c]	
	4	11.225(2)							O,N	93.4(2,1.2)[c]	
										169.6(2,6.1)	
				Mn^{II}O$_6$	O	2.108(5)			O,O	90.0(2,8.3)	
					H$_2$O	2.156(5,13)				171.7(2,4.5)	
						2.219(6,0)					

Table 2. Continued 21.

$[Cu(\mu\text{-obbz})Ni(H_2O)_4].2H_2O$	m	7.764(1)		CuO_2N_2	O	1.905(2,2)	Ni	5.235(1)	O,O	85.2(2)	76a
(red-brown)	$P2_1/n$	12.559(3)	97.56(2)		N	1.923(3,1).			N,N	88.3(1)[c]	
	4	20.083(3)							O,N	93.4(2,2)	
										175.1(1,1.5)	
				$Ni^{II}O_6$	O	2.021(2,10)			O,O	80.03(9)[c]	
					H_2O	2.061(4,37)				91.0(1,6.0)	
										173.2(1,2.0)	
$[Cu(\mu\text{-oxap})Ni(phen)_2](ClO_4)_2$.	tr	12.079(6)	70.91(2)	CuN_4	N	1.918(6,7)	Ni	5.292	N,N	84.2(4,9)[c]	76b
$2H_2O$	P-1	12.409(4)	86.72(4)			1.962(9)				107.3(4)	
(blue red)	2	17.261(8)	89.19(3)			2.054(7)				168.6(4,7)	
				$Ni^{II}N_4O_2$	phenN	2.078(7,24)			N,N	79.7(3,1)[c]	
					O	2.063(5,10)				94.0(2,3.6)	
										173.8(2)	
									O,O	82.6(2)[c]	
									N,O	92.9(3,1.5)	
										171.2(2,5)	
$[Cu(\mu\text{-}Me_2oxpn)Ni(Me_3[10]N_3)]$.	m	13.794(4)		CuN_4	μoxpnN	1.99(2,9)	Ni	5.29	N,N	84.8(10,5)[c]	77
$(ClO_4)_2$	$P2_1$	24.459(7)	95.80(3)							95.0(10,1.1)[c]	
(not given)	4	9.783(3)								175.1(10,7)	
				$Ni^{II}N_3O_2$	μoxpnO	2.01(2,4)			O,O	80.3(7)[c]	
					$Me_3[12]N_3$	2.05(2,1)			N,O	94.0(9,5.9)	
										103.0(8)	
										155.7(8)	
										173.2(8)	
									N,N	92.8(8)[c]	

Table 2. Continued 22.

Compound (colour)	Cryst. cl. / Space group / Z	a, b, c [Å]	α, β, γ [°]	Chromophore		M–L [Å]	M	M…M [Å]		L–M–L [°]	Ref.
[Cu(μ-oxpn)Ni(cth)].(ClO$_4$)$_2$	m	24.704(4)		CuN$_4$	μoxpnN	1.91(3,0)	Ni	5.32	N,N	88.7(8)[c]	77
(not given)	P2$_1$/a	9.802(3)	105.44(3)			2.09(4,9)				94.9(5)[c]	
	4	16.804(3)								88(1,0)	
										175(1,2)	
				Ni^{II}N$_4$O$_2$	cthN	2.13(1,5)			N,N	85.2(4,8)[c]	
					μoxpnO	2.069(9,6)				90.8(5)[c]	
										96.1(5,5.0)	
										174.0(5)	
									O,N	91.3(4,5.5)	
										170.0(4)	
[Cu(μ-oxpn)Ni(phen)$_2$].(ClO$_4$)$_2$.	tr	9.661(3)	86.39(2)	CuN$_4$	μoxpnN	1.99(2,4)	Ni	5.338	N,N	94.6(7,3)[c]	78
MeOH.H$_2$O	P-1	13.522(2)	82.09(2)							85.5(7,1.1)[d]	
(red)	2	17.143(3)	71.73(2)							175.7(8,2.2)	
				Ni^{II}N$_4$O$_2$	phenN	2.08(2,3)			N,N	79.9(7,1.6)[c]	
					μoxpnO	2.08(3,0)				96.1(7,4.1)	
										177.1(2)	
									O,O	80.0(1)[c]	
									O,N	92.2(5,3.0)	
										169.6(6,2.1)	
[Cu(μ-obze)Mn(H$_2$O)$_4$].2H$_2$O	m	17.998(1)		CuO$_2$N$_2$	O	1.929(1,58)	Mn	5.422(1)	O,O	95.59(6)	79
(violet)	P2$_1$/n	6.9949(6)	97.40(1)		N	1.917(2,28)			N,N	85.31(7)[c]	
	4	13.784(1)							O,N	82.31(6)[c]	
										96.81(6)[c]	
				Mn^{II}O$_6$	O	2.162(1,26)			O,O	76.20(5)[c]	
					H$_2$O	2.179(2,62)				91.5(1,9.5)	

Table 2. Continued 23.

Compound				Chromophore							Ref.
[Cu(μ-oxpn)Mn(Me_6[14]ane-N_4)].	m	17.781(7)		CuN_4	oxpnN	1.947(4,6)	Mn	5.436	N,N	84.0(2)[c]	80
$(CF_3SO_3)_2$	$P2_1/a$	18.202(5)	105.48(3)			2.017(4,1)				88.5(2)	
(red)	4	12.893(4)								94.0(2,3)[c]	
										174.1(2,8)	
				$Mn^{II}N_4O_2$	N	2.259(4,2)			O,O	76.9(1)[c]	
					oxpnO	2.274(4,2)			N,N	80.5(1,1)[c]	
						2.172(3,4)				86.2(1,2)[c]	
										95.2(1)	
										160.1(2)	
									N,O	88.9(1,1.9)	
										98.4(1)	
										107.8(2,9)	
										167.5(1,3.4)	
[(acpy)Cu(μ-ox)Cr(salen)].2dmf.	m	16.429(6)		CuO_3N_2	oxO	2.059(7)	Cr	5.482(3)	O,O	76.8(3)[c]	81
MeOH	$P2_1/n$	25.303(14)	109.78(2)			2.378(9)				91.0(3,4.0)	
(dark green)	4	10.191(4)			acpyO	1.910(7)			N,N	93.5(4)[c]	
					N	1.987(9,13)			O,N	95.5(3)[c]	
										91.9(4,5.2)	
										108.0(3)	
										154.2(4)[c]	
										173.6(4)	
				$Cr^{III}O_4N_2$	oxO	2.065(9)			O,O	84.1(3,3.0)	
						1.997(7)				93.3(3,2.4)	
					salenO	1.924(8,4)				176.1(3)	
					N	2.031(9,3)			N,N	78.0(4)[c]	
									O,N	88.4(4,2.5)[c]	
										94.8(4,5.0)	
										170.0(4,1.2)	

Table 2. Continued 24.

[(tren)Cu(μ-im)Zn(tren)].$(ClO_4)_3$.	m	8.924(2)		CuN_5	trenN	2.062(17,17)	Zn	5.84	N,N	83.2(6,1.7)[c]	82
MeOH	$P2_1$	11.590(4)	91.40(2)		μimN	1.954(10)				96.8(6,2.9)	
(blue)	2	16.356(6)								118.6(83.4)	
				$Zn^{II}N_5$	trenN	2.076(13,32)			N,N	83.2(5,5)[c]	
					μimN	1.964(10)				96.8(5,3.1)	
										118.6(6,4.2)	
										176.6(4)	
[(pmdt)(ClO_4)Cu(μ-im)Co	m	15.694(4)		CuN_4O	pmdtN	2.037(6,25)	Co	5.859(2)	N,N	85.5(3,1)[c]	83
$(NH_3)_5$].$(ClO_4)_3$	$P2_1/c$	15.771(4)	112.11(2)		μimN	1.954(6)				94.5(3,1)	
(violet)	4	14.112(3)			O_3ClO	2.886(7)				172.5(4,7.4)	
									N,O	92.9(2,4.9)	
				$Co^{II}N_6$	H_3N	1.967(7,10)			N,N	90.0(3,2.0)	
					μimN	1.933(5)				179.1(3,1)	
[(ClO_4)(Me_5dien)Cu(μ-HCOO)Fe	tr	9.016(3)	78.12(3)	CuN_3O_2	dienN	2.033(7,22)	Fe	5.860	N,N	86.5(3,2)[c]	73
(oep)].ClO_4	P-1	15.377(5)	86.30(4)		μHCOO	1.976(6)				160.0(2)	
(red)	2	19.253(5)	76.23(3)		O_3ClO	2.282(7)			O,N	91.6(3,1.6)	
(at 223 K)										101.4(3,3.0)	
										168.8(2)	
									O,O	86.6(2)	
				$Fe^{III}N_4O$	oepN	2.056(4,1)			N,N	88.3(2)[c]	
					μHCOO	1.968(5)			O,N	100.1(2,2.6)	
[(tpma)Cu(μ-p-$O_2NC_6H_4OPO_3$)	m	17.904(7)		CuN_4O	tpmaN	1.989(9)	Zn	5.86	N,N	80.5(3,8)[c]	84
Zn(tpma)]$(BPh_4)_2$.H_2O	$P2_1/c$	33.436(5)	110.52(3)			2.075(8,18)				107.3(3)	
(green)	4	14.783(3)				2.146(7)				122.7(4,4)	
					O	1.903(7)					
				$Zn^{II}N_4O$	tpmaN	2.047(9,1)			N,N	79.6(3,1.0)[c]	
						2.080(8,25)				114.4(3,1.4)	
										121.6(3)	

Table 2. Continued 25.

Compound	Crystal class Space group Z	a, b, c	α, β, γ	Chromophore	Ligand	M–L	M	M…M	Angle	L–M–L	Ref.
[Cu(μ-im)(μ-cryptand)Zn] $(ClO_4)_3.2H_2O$ (not given)	m $P2_1/a$ 4	18.179(8) 20.333(9) 15.853(6)	113.07(1)	CuN_5	μimN N	1.95(1) 2.14(1,0) 2.19(1,11)	Zn	5.93	N,N	83.2(5,5)[e] 101.6(5,11.3) 143.7;164.6(5)	85
				$Zn^{II}N_5$	μimN N	1.97(1) 2.125(10,5) 2.185(10,25)			N,N	84.5(5,2.2)[e] 108.9(5,14.6) 123.4;166.1(5)	
[(dtma)Cu(μ-im)Zn(dtma)]. $ClO_4.2.5H_2O$ (blue)	or Pbcn 8	14.018(3) 14.886(3) 25.256(6)		CuN_4O	μimN dtmaN dtmaO	1.980(9) 2.051(11,10) 2.135(10) 2.118(8)	Zn	6.0	N,N N,O	82.7(4,1.0)[e] 97.7(4,2) 142.9(4) 178.3(4) 80.1(3)[e] 103.2(4,4.5)	86
				$Zn^{II}N_4O$	μimN dtmaN dtmaO	2.007(1) 2.087(10,7) 2.194(9) 2.045(7)			N,N N,O	83.5(4)[e] 88.1(4,7.0) 111.8(4,3.7) 167.2(3) 79.5(3)[e] 79.6(4,1.9) 138.5(3)	
[(trien)Cu(μ-im)Zn(trien)]. $(ClO_4)_3.H_2O$ (blue)	tr P-1 2	13.262(2) 13.695(4) 8.88(2)	97.64(2) 91.95(2) 88.69(2)	CuN_5	imN trienN	2.042(6) 2.096(7,8)	Zn	6.081(2)	N,N	82.0(3,7)[e] 94.7(3) 106.9(3,2.4) 145.7(3,7)	87
				$Zn^{II}N_5$	imN trienN	2.003(5) 2.068(9) 2.113(10,3) 2.170(20)			N,N	82.1(7,1.6)[e] 92.7(6) 101.6(3) 110.1(4,1.4) 143.7(5,4.8)	

Table 2. Continued 26.

$[Cu(\mu\text{-moim})Mn(salen)]BPh_4$	tr	14.215(3)	121.20(2)	CuN_3O	N	1.933(11,9)	Mn	not given	N,N	82.0(3,1.0)[c]	88
(red brown)	P-1	15.745(3)	114.28(2)			2.014(7)			N,O	96.3(3)[c]	
	2	13.944(3)	58.49(2)		O	1.884(8)				99.5(3)	
				$Mn^{III}N_3O_2$	salenO	1.873(1,6)			O,O	93.1(2)	
					salenN	1.983(9,2)			N,N	82.2(3)[c]	
					moimN	2.165(5)				95.0(3,8)	
									O,N	94.6(3,4.5)	
$[(det)Cu(\mu\text{-pza})Co(NH_3)_5].(ClO_4)_5$	tr	9.567(4)	86.15(3)	CuN_4	N	1.987(7,1)	Co	not given	N,N	84.8(3,3)[c]	89
(purple brown)	P-1	12.652(5)	82.70(3)		μpzN	2.024(5)				95.1(3,7)	
	2	12.808(4)	84.58(3)							172.8(3,3.4)	
				$Co^{III}N_6$	H_3N	1.965(5,20)			N,N	90.0(2,2.5)	
					μpzN	1.986(5)				178.5(2,1)	
$[(bpy)_2Cu(\mu\text{-oxen})Ni].(ClO_4)_2.$	tr	12.179(1)	97.57(1)	CuN_4O_2	bpyN	2.068(5,12)	Ni	not given	N,N	78.7(2,1)[c]	90
MeOH	P-1	12.298(2)	97.52(1)		μoxenO	2.085(3,4)				97.8(2,1.9)	
(violet)	2	11.476(2)	80.29(2)							173.3(2)	
									O,O	82.0(1)[c]	
									O,N	91.6(2,2.4)	
										168.4(2,1)	
				$Ni^{II}N_4$	μoxenN	1.910(5,5)			N,N	83.5(2,1.1)[c]	
						2.000(7,3)				109.4(2)	
										166.5(2,6)	
$[Cl_2Cu(\mu\text{-bdb})Pd(PPh_3)_2Cl]$	m	14.250(6)		CuN_2Cl_2	N	2.005(40,5)	Pd	not given	N,N	80[c]	91
(red-brown)	$P2_1/n$	20.375(8)	102.95(4)		Cl	2.195(20,15)			Cl,Cl	105	
	4	19.181(7)									
				$Pd^{II}P_2CCl$	Ph_3P	2.33(1,0)			P,P	168	
					C	2.00(5)			C,Cl	173	
					Cl	2.36(1)					

Table 2. Continued 27.

$[Cl_2Cu(\mu\text{-hpctp})PdBr_2]C_6H_6.H_2O$	m	10.570(2)		CuN_3Cl_2	N	1.987(13,7)	Pd	not given	N,N	80.6(4,3)[c]	92
(brown)	$P2_1/c$	28.863(4)	98.71(2)			2.352(18)				160.7(5)	
	4	16.099(1)			Cl	2.315(4,11)			Cl,Cl	143.28(16)	
									N,Cl	92.9(4,4.1)	
										108.4(3,2.5)	
				$Pd^{II}N_2Br_2$	N	2.045(11,19)			N,N	86.9(4)	
					Br	2.384(3,6)			Br,Br	89.57(12)	
									N,Br	91.8(3,1.1)	
										177.9(3,1.5)	
$[Cu(\mu\text{-(dpe)py}_2)PtCl_2]$[d]	tr	19.333(1)	103.28(4)	CuN_4	N	1.96(3)	Pt	not given	N,N	90(1,5.0)[c]	93
(purple)	P-1	27.74(1)	108.29(3)			2.01(2,1)				175.5(10,5)	
	4	12.984(5)	76.47(3)								
				$Pt^{II}N_2Cl_2$	N	1.99(3,3)			N,N	179(1)	
					Cl	2.27(1,3)			Cl,Cl	174.2(4)	
									N,Cl	90.0(8,3.5)	
				CuN_4	N	1.96(3,1)			N,N	90(1,2.0)[c]	
						2.01(3,1)				176.5(1,5)	
				$Pt^{II}N_2Cl_2$	N	1.98(3,2)			N,N	177(1)	
					Cl	2.28(1.1)			Cl,Cl	174.1(4)	
									N,Cl	90(1,3.0)	
$[(C_{10}H_{24}N_4)Cu(\mu\text{-Cl})ZnCl_3]$	m	9.828(1)		CuN_4Cl	N	2.004(2,22)	Zn	not given	N,N	84.2(1,5)[c]	94
(purple)	$P2_1/n$	19.741(2)	112.573(2)		μCl	2.6635		113.14(2)		94.93(8)[c]	
	4	10.259(1)								169.3(1,3.9)	
									N,Cl	88.6(1,7)	
				$Zn^{II}Cl_4$	Cl	2.255(3,7)			Cl,Cl	not given	
					μCl	2.3116					

Table 2. Continued 28.

Compound	Cryst. cl. / Space gr. / Z	a / b / c	β	Chromophore	L	M–L	M…M	L,L	L–M–L	Ref.
[(aet)(diena)Cu(μ-NC)Ni(CN)$_3$].	m	11.097(5)		CuN$_5$O	dienN	1.992(4)	Ni not given		not given	95
2H$_2$O	P2$_1$/n	9.869(5)	104.90(2)			2.046(4,7)				
(blue)	4	17.325(4)			aeN	2.003(4)				
					aeO	2.527(3)				
					μCN	2.474(4)				
				Ni^{II}C$_4$	NC	1.864(4,3)			not given	
					μNC	1.865(4)				
[(terpy)Cu(μ-ox)VO(ox)(H$_2$O)].	m	11.247(4)		CuN$_3$O$_2$	terpyN	1.921(5)	V not given	N,N	80.4(2,3)[c]	96
H$_2$O	P2$_1$/c	12.690(4)	97.47(3)			2.025(5,9)			160.8(2)	
(blue)	4	14.759(5)			μoxO	1.966(4)		O,O	78.9(2)	
						2.297(4)		N,O	99.0(2,11.0)	
									171.0(2)	
				V^{IV}O$_6$	O	1.586(5)		O,O	78.7(2,2.6)[e]	
					H$_2$O	2.016(5)			81.8(2,4)	
					oxO	1.985(4,2)			97.9(2,8.3)	
					μoxO	2.002(4)			165.6(2,10.4)	
						2.264(4)				

a Where more than one chemically equivalent distance or angle is present, the mean value is tabulated. The first number in parentheses is the e.s.d., and the second is the maximum deviation from mean value.

b The coordinated atom or ligand is specified in these columns.

c Six - membered metallocyclic ring.

e Five - membered metallocyclic ring.

f Seven - membered metallocyclic ring.

d Two crystallographically independent molecules.

g Four - membered metallocyclic ring.

Table 2 A. Summary of Cu - M and M - M distances [Å] in heterobinuclear copper(II) compounds.

Cu - M	Distances [ref] M = non transition
- Al(III)	2.778(4), 2.784(4) [33a]
- Zn(II)	2.966(2) [37]; 2.985(2) [39]; 3.186(2) [50]; 5.84 [82]; 5.86 [84]; 5.93 [85]; 6.0 [86]; 6.081(2) [87]
- Mg(II)	3.001(4) [41]; 3.031(5) [45]
- Ca(II)	3.262(2) [51]
- Pb(II)	3.344(3) [52]
- Ba(II)	3.69, 3.73 [65]
	M = transition
- Mo(0)	2.558(1) [31]
- Pt(II)	2.765(3) [32]; 3.97 [70]
- Ni(II)	2.897(2) [33b]; 2.898(2) [33c]; 2.925(1) [34]; 2.931(2) [35]; 2.975(2) [38]; 3.038(2) [46]; 3.440(4), 3.452(4) [57a]; 3.596(1) [62]; 3.632(1) [64]; 3.844 [66]; 4.972(1) [57b]; 5.235(1) [76a]; 5.292 [76b]; 5.29, 5.32 [77]; 5.338 [78]
- V(IV)	2.949(3), 2.989(4) [36]
- Co(III)	2.994(2) [40]; 5.859(2) [83]
- Fe(II)	3.00 [42]
- Co(II)	3.03 [42]; 3.025(2) [43]; 3.060(1) [47]
- Mn(II)	3.122(2) [48a]; 3.123 [48b]; 4.612(2) [74a]; 4.127(3) [72b]; 5.195(1) [75]; 5.422(1) [79]; 5.436 [80]
- Fe(III)	3.20 [42]; 3.401(4) [54]; 3.641(1) [54]; 3.570 [60]; 3.596(2) [63]; 3.924(4), 3.938(4) [69a]; 4.039(1) [71]; 4.047(3) [69a]; 4.430 [73]; 4.94, 4.98 [74b]
- Ir(III)	3.416(5) [55]
- Y(III)	3.491 [57b]
- Mn(III)	3.539 [59]
- Cr(III)	3.86 [59]; 5.482(3) [81]
	M = lanthanoide, actinoide
- Pr(III)	3.355(3) [53]; 3.538(2) [58]
- Gd(III)	3.428(1) [56]
- U(VI)	3.51 [57c]

Table 3. Crystallographic and structural data for heterotrinuclear copper(II) compounds[a].

COMPOUND (colour)	CRYST. CL. SPACE G. Z	a[Å] b[Å] c[Å]	α[°] β[°] γ[°]	CHROMO PHORE	M - L [Å]	M-M[Å] M-L-M[°]	L - M - L [°]	REF.
2Cu + M :								
$[(H_2O)_2Fe\{\{\mu\text{-}(prp)_2en\}Cu\}]$. $(ClO_4)_2$ (not given)	m C2/c 2	10.136(3) 22.22(2) 20.476(10)	 98.43(6)	CuO_2N_2	N[b] not given μO "	Fe 2. 976 O not given	O,O[b] not given	98
				$Fe^{II}O_6$	μO 2.18 O 2.06 H_2O 2.12			
$[(ClO_4)Cu(\mu\text{-}Hapen)VO(H_2O)$ $(\mu\text{-}Hapen)Cu(H_2O]ClO_4.EtNO_2$ (not given)	or $P2_12_12_1$ 4	20.928(9) 20.065(7) 10.794(9)		CuO_3N_2	μO 1.907(8,9) N 1.94(1,1) O_3ClO 2.54(1)	V 3.073(3,59 Cu 4.588(3) O 100.3(3,2)	O,O 84.4(3,1.6) 96.3(3) N,N 89.3(4)[c] O,N 93.2(4,1.3)[d] 97.9(4,5.3) 170.0(4,8)	99
				CuO_3N_2	μO 1.920(8,1) N 1.95(1,1) H_2O 2.27(1)	O 100.3(3,6.0)	O,O 86.1(4,5.6) 100.2(4) N,N 88.3(4)[c] O,N 93.5(4,4)[d] 101.1(4,2.9) 162.3(4,7.0)	
				$V^{IV}O_6$	O 1.589(9) μO 2.000(8,11) 2.349(9) H_2O 2.079(8)		O,O 76.1(3,5.4) 96.2(4,7.5) 161.2(4,8.1)	

Table 3. Continued 2.

[Cu(μ-bcb)Mn(μ-bcb)Cu]$(ClO_4)_2$.	m	14.333(2)		CuO_2N_2	μO	1.876(2,10)	Mn	3.194(1,27	O,O	82.5(1,4)	100
2MeCN	$P2_1/n$	17.338(2)	103.74(2)		N	1.912(3,12)	Cu	3.654(9)	N,N	87.4(1,5)[c]	
(not given)	4	28.533(4)					O	not given	O,N	94.8(1,1.3)[d]	
				$Mn^{II}O_4N_4$	μO	2.279(2,6)				not given	
						2.330(2,12)					
					N	2.311(3,15)					
[(ClO_4){Cu(μ-salen)}$_2$Gd$(H_2O)_3$].	m	18.333(5)		CuO_2N_2	μO	1.93(2,1)	Gd	3.30(1)	O,O	85.8(4)[c]	101
{Cu(salen)}$_2$ 0.5$EtNO_2$	$P2_1/a$	33.233(7)	95.94(5)		N	1.92(3,0)	Cu	5.402(3)	N,N	86.2(5)[c]	
(red)	4	11.864(3)					O	99.4(3,1.6)	O,N	94.0(4,6)[d]	
				CuO_3N_2	μO	1.93(4,1)	Gd	3.31(1)	O,O	89.5(3,2.9)	
					N	1.92(4,1)	O	101.1(3,1.3)	N,N	84.7(2)[c]	
					O_3ClO	2.48(5)			O,N	94.1(4,8)[d]	
										96.0(4,7.0)	
				$Gd^{III}O_7$	μO	2.37(2,7)			O,O	71.4(3,4.9)	
					H_2O	2.37(3,3)				95.7(3,9.6)	
				CuO_2N_2	O	1.90(3,2)			O,O	90.1(4,5)	
					N	1.95(4,2)			N,N	86.0(5,5)[c]	
									O,N	92.4(4,9)[c]	
[(O_2NO)Cu(μ-saltn)Gd(H_2O)).	tr	15.694(4)	95.35(0)	CuO_2N_2	μO	1.93(3,10)	Gd	3.37(1)	O,O	83.6(2)	102
$(NO_3)_2$(μ-saltn)Cu].2$EtNO_2$	P-1	14.887(6)	103.18(5)		N	1.955(10,5)	Cu	3.73(1)	N,N	95.1(3)[d]	
(green)	2	10.574(4)	106.95(0)				O	not given	O,N	93.1(3,1.0)[d]	
				CuO_3N_2	μO	1.97(3,0)	Gd	3.35(1)	O,O	84.1(2)	
					N	1.975(30,5)	O	not given		95.6(2)	
					O_2NO	2.40(2)			N,N	95.0(2)[d]	
									O,N	93.1(3,1.0)[d]	
										90.2(2,8)	
				$Gd^{III}O_9$	μO	2.38(3,0)			O,O	50.6(2)	
						2.47(5,6)				139.3(1)	
					H_2O	2.39(1)					
					NO_3,O	2.51(7,4)					

Table 3. Continued 3.

[{Cu(μ-mc)}$_2$Pb](ClO$_4$)$_2$.dmf (not given)	m P2$_1$/c 4	21.239(3) 19.241(1) 13.999(1)	103.47(8)	CuO$_2$N$_2$	μO N	1.94(1,3) 1.97(1,2)	Pb Cu O	3.433(2,9) 6.0 not given	O,O N,N O,N	81.7(4,1.1) 97.1(5,1.3)^d 90.4(5,8)^d 170.8(5,1.5)	103
				Pb^{II}O$_4$N$_4$	μO N	2.53(1,4) 2.66(1,1) 2.68(2,5)			O,O N,N O,N	58.5(3,1.8) 77.6 128.6(3,1.4) 172.4(3) 72.6(4,1.4) 93.0 138.7(4,5.4) 67.5(4,2.9) 81.3(3,3.4) 103.8(4,6.9) 154.5(4,5)	
[Cu$_2$(μ-trf)Ba(H$_2$O)(CF$_3$SO$_3$)]. CF$_3$SO$_3$.H$_2$O (green)	or Pbca 8	28.589(9) 27.786(4) 13.383(3)		CuO$_2$N$_2$	μO N	1.91(-,1) 1.93(-,5)	Ba Cu	3.77(-,8) 3.50		not given	104
				Ba^{II}O$_{12}$	μO trf,O H$_2$O	2.75-3.11 2.89 3.01					
[Cu(μ-saltn)(μ-ac)Zn(μ-saltn). (μ-ac)Cu] (yellow-green)	m P2$_1$/n 2	16.228(5) 11.97(2) 9.239(3)	100.15(3)	CuO$_3$N$_2$	μO N μac,O	1.964(10,0) 1.985(12,4) 2.165(9)	Zn O	3.038(2) 96.2(4,5)	O,O N,N O,N	82.9(4) 95.0(4,33)^c 95.3(5)^d 90.2(4,7)^d 96.2(4,4.4)	105
				Zn^{II}O$_6$	μO μac,O	2.112(9,15) 2.102(9,10)			O,O	75.9(3) 90.2(4,1.5)	
(moe)$_2$Cu(μ-thd)$_3$Ca(moe) (purple)	m P2$_1$/c 4	24.912(8) 10.108(3) 21.897(3)	110.05(2)	CuO$_4$	μO O	1.925 2.376(4)	Ca Ca	3.155(2) 3.318(2)		not given	106
				Ca^{II}O$_8$	O	2.458				not given	

Table 3. Continued 4.

Compound	System / Space group / Z	a, b, c	α, β, γ	Chromophore	M–L	M···M, M–L–M	L–M–L		Ref.
$(moe)_2Cu_2(\mu\text{-thd})_3Sr(moe)$ (purple)	tr P-1 2	12.558(1) 14.950(7) 15.781(6)	73.49(3) 85.69(3) 64.93(3)	CuO_4	μO 1.932 O 2.324(4)	Sr 3.259(2) Sr 3.394(2)		not given	106
				$Sr^{II}O_8$	O 2.478			not given	
$[thd_2Cu_2(\mu\text{-mol})_3Y(hfac)_2]$ (dark blue) (at 238K)	tr P-1 2	14.179(4) 14.425(4) 14.489(4)	88.95(2) 71.45(2) 77.56(2)	CuO_4	μO 1.925(5) μ_3O 1.963(5) thd,O 1.891(5,3)	Y 3.286(2) Cu 3.071(2) μO 103.2(2) μ_3O 93.9(2)	O,O	78.0(2) 93.4(2)[d] 94.4(2,1.4) 171.1(2,2.2)	107
				CuO_5	μO 1.897(6) μ_3O 2.006(4) O 2.439(5) thd,O 1.907(6,4)	Y 3.265(2) μO 104.8(2) μ_3O 96.7(2,4.6)	O,O	78.7(2)[c] 86.4(2,3.1) 93.8(30[d] 95.7(3,1.5) 168.6(2,4.8)	
				$Y^{III}O_8$	μO 2.238(5,20) μ_3O 2.505(6) O 2.738(6,51) hfac,O 2.358(6,31)		O,O	63.7(2,2)[c] 67.0(2,5.4) 73.2(2,1.0)[d] 74.0(2,7.4) 96.0(2,3.3) 135.9(2,10.7)	
$(hfac)Cu_2(\mu\text{-ac})_2(\mu\text{-bdmap})_3Nd.$ (F_3ac) (blue)	m $P2_1/n$ 4	12.027(4) 15.222(4) 32.07(2)	100.13(4)	CuO_3N	μ_3O 1.97(1,1) μac,O 1.915(10,5) O 1.95(2) N 2.04(2,2)	Nd 3.536(3,11 Cu 3.165(5) μ_3O 106.8(7)	O,O O,N	91.2(6,4.9) 174.6(8,4) 88.5(7,1.7)	53
				NdO_8N	μ_3O 2.75(1) μO 2.40(2,4) O 2.52(2,3) N 2.83(2)	Nd 95.5(5,2) 109.6(7,1.0)	O,O O,N	51.5-154.5 66.9-133.9	

Table 3. Continued 5.

$Cu_2(\mu\text{-bdmap})_3(\mu\text{-}F_3ac)_2La(F_3ac)_2$	or	24.95(1)		CuO_3N	μ_3O	1.98(1)	La	3.632(4)	O,O	90.2(7,3.8)	108
(blue)	Pnma	14.806(4)			μO	1.91(1)	Cu	3.265(6)		170.4(6)	
	4	13.724(4)			N	2.06(1)	μ_3O	111.3(9)	O,N	89.2(8,2.3)	
					μF_3ac,O	2.01(1)					
				$La^{III}O_7N_2$	μ_3O	2.84(2)	μ_3O	96,1(6)	O,O	49.5-155.2(3)	
					μO	2.42(1)	μO	113.4(7)	N,N	111.7(8)	
					N	2.85(2)			O,N	61.1-149.3(5)	
					F_3ac,O	2.56(2,11)					
$[\{(Me_6tren)Cu(\mu HCOO)\}_2$.	or	12.517(6)		CuN_4O	N	2.00(1)	Fe	5.422	N,N	84.8(5,3)[d]	73
$Fe(oep)](ClO_4)_2.SbF_6$	Pbcn	29.45(1)				2.15(1,2)				119.2(5,2.9)	
(violet)	4	21.569(8)			O	1.923(9)			N,O	86.2(5)	
										99.9(5,2.9)	
										170.2(5)	
				$Fe^{III}N_4O_2$	N	2.05(1,1)			N,N	90.0(4,6)	
					O	2.103(9,0)			N,O	90.0(4,2.0)	
									O,O	180	
$[\{Cu(acpa)\}_2(\mu\text{-}CrO_4)]$.4MeOH.	m	8.349(2)		CuO_2N_2	μO	1.919(4)	Cr	3.396(9)	O,O	88.8(2)	109
$4H_2O$	$P2_1/m$	17.616(3)	107.40(2)		acpa,O	1.903(4)	O	141.3(2)	N,N	83.8(2)[c]	
(dark green)	2	10.473(3)			N	1.921(4)	Cu	6.443(1)	O,N	92.8(2)	
(at 153K)						1.984(5)				94.6(2)	
										177.2(2,7)	
				$Cr^{VI}O_4$	μO	1.679(4,0)			O,O	109.4(4,1.3)	
					O	1.610(8,4)					
$[\{Cu(acpa)\}_2(\mu\text{-}MoO_4)].4H_2O$	m	8.486(2)		CuO_2N_2	μO	1.929(6)	Mo	3.576(1)	O,O	87.4(3)	109
(dark blue)	$P2_1/m$	18.043(3)	95.82(1)		acpa,O	1.907(6)		149.0(4)	N,N	83.6(3)[c]	
(at 213K)	2	9.753(2)			N	1.917(8)	Cu	6.703(2)	O,N	93.8(3)	
						1.979(7)				95.1(3)[d]	
										177.3(3,1.3)	
				$Mo^{VI}O_4$	μO	1.782(6,0)			O,O	109.6(5,1.6)	
					O	1.730(9)					
						1.73(1)					

Table 3. Continued 6.

[{(Me$_6$tren)Cu(μ-NC)}$_2$Fe(oep)].	or	13.093(1)		CuN_5	N	2.12(1,13)	Fe	4.99	N,N	85.6(4,4)[c]	74b
$(SbF_6)_3$	Pbcn	28.964(8)			μCN	1.94(1)				94.8(4,1.2)	
(dark violet)	4	21.243(4)								119.4(4,1.0)	
										178.8(4)	
				$Fe^{III}N_4C_2$	N	1.995(8,1)			N,N	90.0(3,4)[d]	
					μNC	1.94(1)					
[{(tren)Cu(μ-NC)}$_2$Fe(CN)$_4$].	m	13.582(1)		CuN_5	N	2.048(2,2)	Fe	not given	N,N	83.8(1,6)[c]	110
$12H_2O$(dark blue)	$P2_1/c$	10.112(1)	95.33(1)			2.167(2)				101.6(1,10.1)	
	2	13.940(1)			μCN	1.958(2)				140.7(1)	
										169.4(1)	
				$Fe^{II}C_6$	μNC	1.894(2,0)			C,C	89.9(1,4.1)	
					NC	1.930(2,6)				180.0(1)	
[{(H$_2$O)(dien)Cu}$_2$(μ-im) .	m	37.295(7)		CuN_4O	μim,N	1.951(5)	Co	5.853	N,N	83.9(3,4)[c]	111
Co(NH$_3$)$_4$](NO$_3$)$_5$	C2/c	7.360(1)	105.65(3)		dien,N	2.024(10,6)	Cu	11.705		96.5(3,1.8)	
(purple brown)	4	13.610(3)			H_2O	2.409(7)				155.3(3)	
										176.7(3)	
									O,N	94.6(,7.6)	
				$Co^{III}N_6$	μim,N	1.952(6,10)			N,N	90.0(3,1.9)	
					H_3N	1.962(6,0)				180.0(1)	

Table 3. Continued 7.

$[Cu_2(\mu\text{-tem})Zn](ClO_4)_3.3H_2O$	tr	14.262(4)	117.22(2)	CuO_2N_2	μO	1.902(6,20)	Zn	3.08	O,O	81.5(3)	112
(dark green)	P-1	15.745(4)	111.03(2)		N	1.97(1,2)	Cu	8.13	N,N	86.2(5)[c]	
	2	16.072(4)	90.28(2)				O	101.2(3,3.7)		96.8(4)	
									O,N	93.6(3,1.0)[d]	
										96.8(4)	
										170.2(11,7.3)	
				CuO_2N_2	O	1.931(6,10)	Zn	5.61	O,O	89.7(3)	
					N	2.009(8,2)			N,N	84.9(3)[c]	
									O,N	92.6(3,8)[d]	
										175.5(3,1.3)	
				$Zn^{II}N_3O_2$	N	2.108(8,38)			N,N	85.5(3)[c]	
					μO	2.007(7)				111.1(3,8.0)	
						2.174(6)			N,O	88.7(3,1.4)[d]	
										102.2(3,9)	
										148.4(3,11.5)	
$[\{Cu(\mu\text{-bado})\}_2Ni(H_2O)_2](ClO_4)_2.$	tr	12.080(1)	101.97(1)	CuN_4	N	1.958(6,11)	Ni	5.316(1,1)	N,N	83.4(2,2)[c]	113
$4H_2O$	P-1	13.054(1)	90.82(1)			2.0002(7,10)				88.0(2,4)[d]	
(ruby violet)	2	14.775(2)	101.88(1)							174.8(2,3.6))	
				$Ni^{II}O_6$	O	2.061(5,23)			O,O	81.5(2,1)[c]	
					H_2O	2.077(,8)				92.0(2.4.1)	
										174.2(2,3.1)	
$[\{Cu(\mu\text{-bado})\}_2Gd(H_2O)_4].$	m	11.358(5)		CuN_4	N	1.98(1,4)	Gd	5.683(1,2)	N,N	83,3(5,8)c	114
$(NO_3)_3.2H_2O$	$P2_1/c$	16.932(4)	103.47(3)				Cu	7.681(1)		93.5(5,9)[d]	
(ruby violet)	4	23.625(2)								90.2(7,9)	
										172.8(7,1.6)	
				$Gd^{III}O_8$	O	2.364(10,66)			O,O	75.9(5,10.7)	
					H_2O	2.422(10,6)				115.7(6,4.6)	
										139.3(5,7.0)	
										152.3(4,9)	

Table 3. Continued 8.

Cu + 2M:

$[\{(NH_3)_4Co(\mu\text{-}OH)_2Cu\}(SO_4)_2$. $4H_2O$ (pale pink)	m $P2_1/c$ 2	6.220(1) 16.956(1) 9.994(2)	 95.44(1)	CuO_4	μHO	1.936(3,10)	Co 2.888(1,0) O 97.O(1,4)	O,O	82.2(1,0)	115
				$Co^{III}N_4O_2$	H_3N μHO	1.952(4,9) 1.921(3,3)		N,N O,O N,O	90.5(2,2.1) 178.2(2) 83.6(1) 90.8(2,4.6) 174.7(2,1.5)	
$[\{(en)_2Co(\mu\text{-}OH)_2\}_2Cu(SO_4)]$. $SO_4.2H_2O$ (red brown)	m $P2_2$ 2	6.262(3) 13.720(3) 14.301(4)	 101.46(3)	CuO_5	μHO O_3SO	1.923(4) 1.963(4,11) 2.516(4)	Co 2.898(1,4) O 97.0(2,1.0)	O,O	83.0(2,2.4) 96.2(2,4.8) 174.9(1,1)	115
				$Co^{III}N_4O_2$	en,N μHO	1.950(4,33) 1.915(5,17)		N,N O,O O,N	86.1(2,9)[c] 92.4(2,2.0) 177.2(2,1) 83.8(2,9) 91.3(2,3.2) 172.8(2,1.4)	
$[\{(bispictn)Cr(\mu\text{-}OH)_2\}_2$. $Cu\ (ClO_4)](ClO_4)_3.2H_2O$ (blue violet)	m $P2_1/c$ 4	14.650(11) 14.226(8) 22.754(14)	 108.32(6)	CuO_5	μHO O_3ClO	1.912(4,9) 2.503	Cr 2.899(2,4) O 98.8(3,8)	O,O	80.8(2,1) 99.0(2,6) 175.3(2,1.3)	116
				$Cr^{III}N_4O_2$	N μHO	2.041(5,17) 1.906(4,9)		N,N O,O N,O	80.9(2,1.0)[c] 96.3(2,2)[d] 94.1(2,7) 81.1(2,1) 93.0(2,3.1) 171.9(2,2.6)	

Table 3. Continued 9.

[{(cyclam)Cr(μ-OH)$_2$}$_2$	m	8.770(1)		CuO_6	μHO	1.930(3,9)	Cr	2.958(1)	O,O	80.8(1)	116
$Cu(ClO_4)_2$]$(ClO_4)_2 4H_2O$	$P2_2/c$	20.791(3)	99.47(1)		O_3ClO	2.838(3,0)	O	99.5(1,7)		99.2(1)	
	2	11.833(2)									
				$Cr^{III}N_4O_2$	N	2.082(5,6)			N,N	82.6(2,2)[c]	
					μHO	1.945(4,9)				91.0(2,4)[d]	
										97.7(2)	
									O,O	80.1(1)	
									N,O	92.9(2,2.8)	
										170.2(2,1.0)	
[{$(OC)_4$Re(μ-MeCO)$_2$}$_2$Cu]	m	11.2231(9)		CuO_4	μMeCo	1.920(3.0)	Re	3.617(0)	O,O	90.0(2,2.9)	117
(deep red)	$P2_1/c$	7.3648(5)	90.791(6)								
	2	23.9179(10)									
				ReC_6	μMeOC,C	1.977(6,19)			C,C	90.0(4,5.9)	
					μMeOC,C	2.152(5,3)				175.6(4,1.3)	
[{Me_2Ga(μ-pz)$_2$}$_2$Cu]	m	8.522(3)		CuN_4	N	1.987(4,7)	Ga	3.501	N,N	90.0(2,2.6)	118
(lilac)	$P2_1/c$	18.091(5)	105.90(4)								
	2	7.463(3)									
				$Ga^{III}N_2C_2$	N	1.975(4,5)			N,N	92.6(2)	
					Me,C	1.962(8,5)			C,C	127.6(3)	
									N,C	107.8(3,2.0)	
[{Me_2Ga(μ-Me_2pz)$_2$}$_2$Cu]	m	16.645(4)		CuN_4	N	1.959(4,6)	Ga	3.790(1,10	N,N	109.5(2,5.8)	118
(dark red)	I2/a	12.964(1)	104.69(2)								
	4	14.344(4)									
				$Ga^{III}N_2C_2$	N	1.996(4,3)			N,N	103.3(2,4)	
					Me,C	1.973(10,8)			C,C	126.7(4,5)	
									N,C	106.4(3,3)	

Table 3. Continued 10.

[{$(NH_3)_2Pt(\mu$-$C_5H_4NO)_2$}$_2$Cu]. $(NO_3)_2$ (orange red)	or Fdd2 8	23.281(4) 22.551(4) 10.852(2)		CuO_4	O	1.960(9,9)	Pt	2.6312(4)	O,O	≈90 162.1(3)	119
				$Pt^{II}N_4$	N H_3N	2.025(9,9) 2.03(1,1)			N,N	90.0(5,1.3) 174.6(4,4.1)	
[{$(MeNH_2)_2Pt(\mu$-$C_5H_4NO)_2$}$_2$Cu]. $(NO_3)_2$ (orange red)	m $P2_1/n$ 4	11.448(4) 18.771(2(16.085(3)	105.8(2)	CuO_4	O	1.974(4,9)	Pt	2.638(1,6)	O,O	90.0(2,3.3) 163.7(4,5)	119
				$Pt^{II}N_4$	N MeH_2N	2.028(5,5) 2.058(5,4)			N,N	90.0(2,4.2) 176.9(2,2.7)	
[{$(NH_3)_2Pt(\mu$-meu)(μ-1-mec)}$_2$. Cu]$(NO_3)_4.6H_2O$ (green)	tr P-1 1	11.522(6) 10.924(4) 10.736(2)	91.51(3) 109.08(3) 114.43(3)	CuO_4	μ-meu,O μ-mec,O	1.931(12) 1.988(9)	Pt	2.681(1)	O,O	92.3(4)	120
				$Pt^{II}N_4$	μ-meu,N μ-mec,N H_3N	2.045(13) 2.036(11) 2.075(14,7)			N,N	90.0(5,1.6) 178.9(5,1)	
[{$(NH_3)_2Pt(\mu$-meu)$_2$}$_2$Cu] SO_4. $8H_2O$ (green)	tg $I4_1/a$ 8	20.470(7) 23.509(8)		CuO_4	O	1.936(12,7)	Pt	2.685(1)	O,O	90.0(6)	121
				$Pt^{II}N_4$	μ-meu,N H_3N	2.016(14,4) 2.046(1,1)			N,N	90.0(6,8.3) 178.3(6,7)	
[{$(NH_3)_2Pt(\mu$-NHCOMe)$_2$}$_2$Cu]. $(ClO_4)_2$ (green)	m $P2_1/n$ 2	9.924(1) 7.364(1) 16.589(2)	91.08(1)	CuO_4	O	1.949(3,0) 1.999(4,0)	Pt	2.705(2,0)	O,O	90.1(2,0)	122
				$Pt^{II}N_4$	N H_3N	2.003(5,8) 2.050(5,1)			N,N	90.0(2,3.6) 177.4(2,1.4)	

Table 3. Continued 11.

[{(tmeda)Pt(μ-1-meu)$_2$}$_2$Cu].	tr	10.127(1)	68.54(1)	CuO_4	O	1.932(3,4)	Pt	2.9843(1)	O,O	90.0(1,1.1)	123
$(ClO_4)_2$	P-1	10.816(2)	65.75(1)								
(green)	1	10.317(2)	89.52(2)								
				$Pt^{II}N_4$	N	2.067(3,35)			N,N	84.2(1)[c]	
										91.9(1,5.2)	
										176.4(1,6)	
[{(C_6Me_6)Ru(μ-Cl)(μ-(MeO)$_2$	m	19.136(1)		CuO_4Cl_2	O	1.948(4,17)	Ru	not given	O,O	90.0(2,9)	124
PO)$_2$}$_2$Cu].2CH_2Cl_2	$P2_1/n$	15.157(1)	92.91(3)		μCl	2.881(2,0)	Cl	93.6(1)		180.0(0)	
(not given)	2	8.884(1)							O,Cl	88.8(1,9.3)	
				$Ru^{II}C_6P_2Cl$	C	2.266(7,45)			C,C	78.4(2)	
					P	2.269(2,6)			P,P	84.02(7)	
					μCl	2.392(2)			C,P	100.7(2,8.9)	
										130.9(2)	
										166.3(2)	
									P,Cl	85.8(1,1.8)	
[Cu(μ-paba)$_6$Fe$_2$].2CH_2Cl_2	m	14.538(7)		CuO_6	O	1.973(4,0)	Fe	3.483(1)	O,O	90.8(2,3.3)	125
(dark)	$P2_1/c$	13.019(8)	108.89(4)			2.014(6,0)				180.0(1)	
	2	22.673(11)				2.412(7,0)					
				$Fe^{II}N_6$	N	1.894(7)			N,N	78.5(3,6)[5]	
						1.923(6)				92.2(3,3.7)	
						1.943(7,2)				101.2(3,1.2)	
						1.960(6,3)				167.0(3,3)	
[{(tmac)Mn}$_2$(μ-niox)$_3$Cu].	m	30.39(2)		CuN_6	N	1.96	Mn	3.54		not given	126
(ClO_4).2MeCN	C2/c	9.518(11)	113.46(6)			2.00					
(dark)	4	22.53(2)				2.36					
(at 173K)											
				$Mn^{IV}O_3N_3$	O	1.87	Mn	7.06		not given	
					tmac,N	2.08					

Table 3. Continued 12.

[{(tmac)Fe}$_2$(μ-dmg)$_3$Cu]. (ClO$_4$)$_2$.0.5MeOH (not given)	m C2/c 4	29.144(5) 12.330(3) 14.860(2)	119.22(4)	CuN_6	N	2.022(5,47) 2.324(4) 2.523(6)	Fe	3.58	N,N	64.1(2) 78.4(3,6)[c] 80.5-160.3(2)	127
				$Fe^{III}O_3N_3$	O tmac,N	1.928(5,12) 2.215(6,20)	Fe	7.0		not given	
{(MeOH)$_2$Li}$_2$(μ-tdp)$_2$Cu (brown)	m P2$_1$/n 4	11.676(21) 15.899(24) 16.726(23)	100.63(12)	CuO_4S_2	μO S	1.929 2.222 2.482	Li O	not given 91.4(-,4.1)	O,O S,S O,S	88.0(-,8.9) 176.9 113.6 83.1(-,8)[c] 97.4	128
				$Li^{I}O_4$	μO MeOH	1.948 "			O,O	96.7(-,7.6) 125.5(-,4.8)	
[{(pr^{i}O)$_4$Ti$_2$(μ-pr^{i}O)}(μ$_3$-pr^{i}O)$_2$. (μ-pr^{i}O)$_2$Cl] (green)	tr P-1 2	10.085(2) 10.210(2) 21.551(4)	84.40(3) 84.03(3) 60.86(3)	CuO_4Cl	μ$_3$O μO Cl	2.153(2,4) 2.008(2,1) 2.1720(12)	Ti μ$_3$O μO	2.957(-,2) 87.4(1,8) 96.6(1,0)	O,O O,Cl	73.5(1,3.1) 146.3(1) 106.8(1,7) 144.(1,1.3)	129
				$Ti^{IV}O_6$	μ$_3$O μO O	2.124(2,36) 1.982(2,1) 2.076(2,11) 1.778(2,6)	Ti μ$_3$O μO	3.076 90.8(1,1) 95.6(1)	O,O	74.6(1,40) 98.9(1,5.8) 143.2(1,4) 166.6(1,1)	
[{(pr^{i}O)$_4$Zr$_2$(μ-pr^{i}O)}(μ$_3$-pr^{i}O)$_2$. (μ$_3$-pr^{i}O)$_2$CuCl] (green) (at 118K)	m P2$_1$/n 4	9.459(2) 24.351(4) 16.654(3)	95.06(1)	CuO_4Cl	μ$_3$O μO Cl	2.197(6,11) 1.999(6,1) 2.164(3)	Zr μ$_3$O μO	3.073(2,4) 87.7(2,6) 97.1(2,1)	O,O O,Cl	76.7(2,5.5) 150.52(23) 104.7(2,1.5) 144.4(2,1.3)	130
				$Zr^{IV}O_6$	μ$_3$O μO O	2.240(6,19) 2.099(6,7) 2.192(6,16) 1.926(6,12)	Zr μ$_3$O μO	3.3102(13) 95.3(2,2) 98.07(22)	O,O	102.5(3,2.0) 138.72(23)	

Table 3. Continued 13.

[{(pr^{i}O)$_2$Ba$_2$(μ-η^2-thd)}(μ$_3$-η^2-	tr	10.364(6)	97.05(1)	CuO$_4$N	μ$_3$O	1.925(10,5)	Ba	3.325(2)	O,O	84.7(4,2.8)	131
OCHMeCH$_2$NMe$_2$)$_2$(μ-thd)Cu]	P-1	14.213(3)	97.96(3)		μO	2.28(1)		3.520(2)		96.2(4,4.3)	
(turquoise blue)	2	26.570(5)	108.95(3)		O	1.92(1)	μ$_3$O	93.1(4,4.8)	O,N	85.1(5)^c	
(at 173K)					N	2.05(1)	μO	82.5(3)		95.4(5,1.1)	
										165.4(5)	
				Ba^{II}O$_6$N	O	2.60(1)	Ba	4.074(2)	O,O	54.8(-)	
					μO	2.70(1,4)	μ$_3$O	91.6(4,2.9)		142.9(3)	
					μ$_3$O	2.74(1,1)	μO	94.5(4)	O,N	62.4(-)	
					N	2.85(1)				143.4(3)	
				Ba^{II}O$_8$	O	2.74(1,7)			O,O	55.3(-)	
					μO	2.87(1)				162.7(5)	
					μ$_3$O	2.72(1,2)					
[{Ni(μ-edta)}$_2$Cu(en)$_2$].	m	8.468(1)		CuN$_4$O	en,N	2.040(3,4)	Ni	3.788	N,N	84.56)13)^c	132
[Cu(en)$_2$(H$_2$O)$_2$].2H$_2$O	P2$_1$/c	21.752(2)	104.79(1)		μedta,μO	2.376(2,0)	O	117.4(1)	O,N	90.5(1,4)	
(not given)	4	12.598(1)									
				Ni^{II}O$_4$N$_2$	μedta,N	2.053(3,17)			O,O	90.5(1,4.6)	
					μedta,μO	2.052(3)				111.69(9)	
					O	2.054(3,16)			N,N	88.05(11)^c	
									O,N	82.2(1,2.2)^c	
										95.0(1,1.7)	
										164.9(1,1)	
				CuN$_4$O$_2$	en,N	2.010(3,1)			N,N	85.80(13)^c	
					H$_2$O	2.635(3,0)			O,N	87.2(1,7)	
[{(H$_2$O)$_2$Mn(μ-Cl$_2$opba)}$_2$	tr	7.066(7)	105.64(2)	CuN$_4$O$_2$	en,N	2.016(7,4)	Mn	3.836(1)	N,N	85.3(3)^c	133
Cu(en)$_2$].2dmso	P-1	11.844(1)	97.67(5)		μO	2.454(6,0)	O	not given	O,O	180	
(brown)	1	14.22(5)	102.13(3)								
				Mn^{II}O$_4$N$_2$	μO	1.973(5)	Mn	7.671(2)	O,O	112.2(2)	
					O	1.961(5)				160.0(2)	
					N	1.918(7,7)			N,N	82.2(3)^c	
					H$_2$O	2.250(5,12)					

Table 3. Continued 14.

Compound	Crystal data	a, b, c (Å)	α, β, γ (°)	Chromophore	M–L (Å)	M–L–M (°)	L–M–L	(°)	Ref.
$[(py)_4Cu(\mu\text{-}O)_2Nb_2F_{10}][pyH]$ (deep blue) (at 153K)	tg $I4_122$ 4	11.408(3) 30.36(1)		CuN_4O_2	py,N 2.06(1,0) μO 2.417(9,0)	Nb not given	N,N	90.0	134
				Nb^VF_5O	F 1.932(6,5) 2.099(8) μO 1.728(8)		F,F F,O	85.9(2,3.6) 168.3(3,4) 95.9(2,2) 180	
$[(py)_4Cu(\mu\text{-}O)(\mu\text{-}F)Nb_2F_4]$ (deep blue) (at 173K)	m C2/c 4	10.561(3) 13.546(6) 16.103(4)	97.77(2)	CuN_4OF	py,N 2.040(2,0) 2.080(2,0) μO 2.3630(2) μF 2.3630(2)	Nb not given	N,N	88.08(6)	134
				Nb^VF_5O	F 1.930(1,9) μF 1.9340(2) μO 1.9340(2)		F,F	90.00(6,49)	
$[(bpy)_2Cu(\mu\text{-}O)Cr(O_2)(\mu\text{-}O).$ $CrO_3]. 2H_2O$ (not given)	tr P-1 2	7.716(3) 9.656(3) 15.517(5)	77.41(3) 81.04(3) 82.33(3)	CuN_4O	bpy,N 2.00(2,5) μO 2.240(8)	Cr not given O 137.0(5)	N,N N,O	82.0(4,5)[c] 100.8(4,7) 138.9(4) 172.2(3) 86.2(4,1.3) 97.1(3) 124.0(3)	135
				$Cr^{VI}O_4$	O 1.59(2,1) μO 1.617(8) 1.76(2)	O 133.8(6)	O,O	109.5(6,2.8)	
				$Cr^{VI}O_4$	O 1.58(2,1) μO 1.77(1)		O,O	104.5(5,3.1)	

Table 3. Continued 15.

$[(NH_3)_5Co(\mu\text{-}im)Cu(dien)(\mu\text{-}im)$. $Co(NH_3)_5]Cl_6$ (brown)	m C2/c 4	32.414(4) 9.421(1) 15.167(5)	 90.91(2)	CuN_5	dien,N 2.05(1) 2.22(1) μim,N 2.046(3)	Co not given	N,N	84.1(4,5)[c] 150.6(3)	136
				$Co^{III}N_6$	H_3N 1.978(3,22) μim,N 1.921(2)		N,N	90.0(1,2.2) 179.2(1,4)	
$\{(CO)_5Mo(\mu\text{-}dps)\}_2Cu$[e] (dark brown)	tr P1 2	11.317(1) 15.032(1) 18.752(1)	95.613(6) 97.984(6) 90.22(1)	CuO_2N_2	O not given N not given	Mo 7.078(1)	O,N	91.60(6)	137
				Mo^0C_5P	OC 2.019(3,21) P not given				
				CuO_2N_2	O not given N not given	Mo 7.079(1)	O,N	91.77(6)	
				Mo^0C_5P	O 2.020(3,18) P not given				
$[\{(cp)Fe(\mu\text{-}C_5H_4CO_2)\}_2$ $Cu(dmim)_2]$ (green) (at 155K)	tr P-1 2	9.468(2) 9.849(2) 16.011(4)	90.44(2) 94.85(2) 90.47(2)	CuO_2N_2	O 1.972(4,11) N 1.995(5,6)	Fe not given	O,O N,N O,N	91.8(2) 93.7(2) 89.8(@,1) 162.3(2,4.5)	138
				FeC_{10}	cp,C not given				
$[(cp)Fe(\mu\text{-}C_5H_4CO_2)_2Cu(mim)_2]$ (golden brown) (at 155K)	m C2/c 4	21.145(3) 7.858(1) 18.573(3)	 114.61(1)	CuO_2N_2	O 1.985(4,0) N 1.973(4,0)	Fe not given	O,N	89.0(2)	138
				FeC_{10}	cp,C not given				
$[\{(cp)Fe(\mu\text{-}C_5H_4CO_2)\}_2Cu(py)_2]$ (yellow brown)	m $P2_1/c$ 2	14.761(5) 5.922(2) 15.913(6)	 102.69(3)	CuO_4N_2	O 2.123(5,0) 2.304(5) py,N 1.977(5)	Fe not given	O,O O,N	59.4(2)[f] 89.5(2,6)	139
				FeC_{10}	C not given			not given	

Table 3. Continued 16.

[{(cp)Fe(μ-$C_5H_4CO_2$)}$_2$Cu(py)$_2$] (yellow brown)	tr P-1 1	5.986(2) 8.038(2) 15.512(2)	104.42(2) 93.11(2) 99.95(2)	CuO_4N_2	O 1.975(3,0) 2.53(-,0) py,N 1.998(5,0)	Fe not given	O,N	89.5(2)	139
				FeC_{10}	C not given				
[{(cp)Fe(μ-$C_5H_4CO_2$)}$_2$Cu(im)$_2$] (yellow brown)	tr P-1 1	7.475(3) 9.296(3) 10.090(3)	111.05(2) 92.38(4) 101.69(3)	FeO_4N_2	O 1.957(9,0) 2.76(-,0) im,N 1.955(9,0)	Fe not given	O,N	88.8(4)	139
				FeC_{10}	C not given			not given	
{(cp)Fe(μ-hbcpa)}$_2$Cu (dark brown)	m P2$_1$/c 4	12.668(2) 10.422(2) 20.921(3)	91.10(1)	CuO_2N_2	O 1.990(4,11) N 1.973(4,20)	Fe 4.300(1) 4.350(1)	O,O N,N O,N	137.0(2) 138.3(2) 95.2(2,3)[d] 99.7(2,3.5)	140
				FeC_{10}	C 1.646(8,7)		cp,cp	178.9(4,4)	
[{(CO)$_5$W(μ-odp)}$_2$Cu] (green)	m P2$_1$/c 2	15.063(3) 15.891(3) 19.716(4)	106.97(1)	CuO_4	O not given				141
				W^0C_5O	OC not given O not given				

a Where more than one chemically equivalent distance or angle is present the mean value is tabulated. The first number in parenthesis is e.s.d., the second is a maximum deviation from the mean value.
b The chemical identity of coordinated atoms.
c Five-membered metallocyclic ring.
d Six-membered metallocyclic ring.
e There are two crystallographically independent molecules.
f Four-membered metallocyclic ring.

Table 3A. Summary of Cu - M and M - M distances [Å] in heterotrinuclear copper(II) compounds

Cu - M	Distances [ref] M = non transition
- Zn(II)	3.038(2) [105]; 3.08, 5.62 [112]
- Ca(II)	3.155(2), 3.318(2) [106]
- Sr(II)	3.259(II), 3.394(2) [106]
- Ba(II)	3.325(2), 3.520(2) [131]; 3.70, 3.85 [104]
- Pb(II)	3.424(2), 3.442(2) [103]
- Ga(III)	3.501(1), 3.780(1), 3.799(1) [118]
	M = transition
- Pt(II)	2.6312(4), 2.632(1), 2.645(1) [119]; 2.681(1) [120]
	2.685(1) [121]; 2.705(2) [122]; 2.9843(1) [123]
- Co(III)	2.888(1), 2.898(1), 2.899(1) [115]
- Cr(III)	2.899(2), 2.959 [129]
- Ti(IV)	2.955, 2.959 [129]
- Fe(II)	2.976(1) [98]; 3.483(1) [125]
- V(IV)	3.015(3), 3.132(3) [99]
- Zr(IV)	3.069(2), 3.077(2) [130]
- Mn(II)	3.167(1), 3.221(1) [100]
- Y(III)	3.265(2), 3.286(2) [107]
- Cr(VI)	3.3961(9) [109]
- Mn(IV)	3.54 [126]
- Mo(VI)	3.576(1) [109]
- Fe(III)	3.58 [127]; 4.99 [74b]; 5.422 [73]
- Re(0)	3.617 [117]
- Ni(II)	3.788 [132]; 5.315(1), 5.316(1) [113]
- Mn(III	3.836(1) [133]
- Fe(0)	4.300(1), 4.350(1) [140]
- Mo(II)	7.078, 7.079(1) [137]

Table 3A. Continued 2

	M = lanthanoide
- Gd(III)	3.30(1), 3.31(1) [101]; 3.36, 5.34 [101]; 3.35(1), 3.37(1) [102]
- Nd(III)	3.525(3), 3.547(3) [53]; 5.681(1), 5.684(1) [114]
- La(III)	3.632(4) [108]
Cu(II) - Cu(II)	3.071(2) [107]; 3.165(5) [53]; 3.265(6) [108]; 3.50 [104]; 3.645(1) [100]; 3.73(1) [102]; 4.558(3) [99]; 4.561(3) [101]; 5.402(3) [101]; 6.0 [103]; 6.443(1) [109]; 7.681(1) [114]; 8.13 [112]; 11.705 [111]
Ti(IV) - Ti(IV)	3.076 [129]
Zr(IV) - Zr(IV)	3.3102(13) [130]
Ba(III) - Ba(III)	4.074(2) [131]
Fe(III) - Fe(III)	7.0 [27]
Mn(IV) - Mn(IV)	7.06 [126]
Mn(III) - Mn(III)	7.671(2) [133]

Table 4. Crystallographic and structural data for heterotetranuclear copper(II) compounds[a].

COMPOUND (colour)	CRYST. CL. SPACE GR. Z	a[Å] b[Å] c[Å]	α[°] β[°] γ[°]	CHROMO-PHORE	M – L [Å]	M – M [Å] M -L- M[°]	L – M – L [°]	REF.
3Cu + M :								
$[(NO_3)_2Cu_3(\mu\text{-mhp})_6.$	m	18.2394(2)		CuO_2N_2	μO[b] 1.985(30)x2	Mg not given	O,O 78.39(13)	142
$Mg(mhpH)_2]$	$P2_1/n$	14.4374(8)	99.553(5)	(x1)	N 2.001(3)x2	O[b] not given	N,N 95.75(13)	
(bright green)	4	20.7395(2)					O,N 93.86(13)	
							166.56(13)	
				CuO_2N_2	μO[b] 1.936(30)		O,O[b] 157.41(13)	
				(x2)	N 2.007(30,9)		N,N 153.03(13)	
					O_2NO 2.020(30)		O,N 92.0(13,1.7)	
				MgO_6	μO 2.132(30)x2		O,O 72.85(13)	
					O 2.042(30)x2		89.8(13,3.8)	
					2.064(30)x2		178.28(13)	
$[\{(thd)Cu\}_3(\mu_3\text{-mol})_2(\mu\text{-mol})_3.$	m	33.084(5)		CuO_5	μ_3O 1.988(8)	Y 3.204(2)	O,O 81.9(3)	107
Y(thd)]	Cc	13.980(2)	121.70(10)		μO 1.916(7)	Cu 3.107(2)	92.0(3)[c]	
(dark blue)	4	18.789(3)			O 2.484(11)	μ_3O 97.3(3,5.9)	93.0(3,1.7)	
(at 283 K)					thdO 1.925(9,22)	μO 99.7(4)	173.9(5,2.8)	
				CuO_5	μ_3O 1.977(9)	Y 3.246(2)	O,O 79.2(3,1.6)	
					2.407(8)	Cu 3.711(2)	92.1(4)[c]	
					μO 1.956(6)	μ_3O 88.9(4,4.5)	94.8(4,1.1)	
					thdO 1.908(10,12)	118.0(4)	103.2(4,2.3)	
						μO 98.6(3)	171.2(4,7)	
				CuO_5	μ_3O 1.912(8)	Y 3.377(2)	O,O 91.4(4,5.8)	
					μO 1.943(9)	μ_3O 101.6(4)	92.1(4)[c]	
					O 2.588(11)	μO 105.4(4)	160.3(4,3.6)	
					thdO 1.937(11,3)			
				$Y^{III}O_8$	μ_3O 2.455(8,20)		O,O 65.2(3)[d]	
					μO 2.285(10,21)		62.5(-)	
					O 2.530(11)		156.5(2)	
					thdO 2.327(7,22)			

Table 4. Continued 2

Compound	Cryst. cl. / Space gr. / Z	a, b, c	α, β, γ	Chromophore	M–L	M...M / M–L–M	L–M–L	Ref.
[(MeOH)Cu_3(μ-OMe)(μ-chp)$_5$.	tr	11.2055(19)	81.345(13)	CuO_2N_2	μMeO 1.914(3)	Cu 2.9699(7)	O,O 79.25(12)	143
La$(NO_3)_3$(MeOH)].1.5MeOH.	P-1	12.9050(28)	85.310(11)		μO 1.976(3)	μMeO 109.18(14)	N,N 98.95(14)	
0.5Et_2O	2	17.059(4)	76.025(13)		N 1.980(4,21)	O 94.6(11)	O,N 92.1(1,6.8)	
							164.4(1,3.6)	
				CuO_4N	μMeO 1.899(3)		O,O 73.0(11)	
					μO 2.246(3)		90.6(1,4.9)	
					O 1.916(3)		165.95(12)	
					N 2.053(3)		O,N 94.5(1,1.9)	
					MeHO 2.050(3)		135.4(1.3.4)	
				CuO_2N_2	μO 1.958(3,4)	La 3.719(1)	O,O 80.92(11)	
					N 1.986(3,4)	O 108.2(1,6)	N,N 94.94(14)	
							O,N 93.5(1,1.4)	
							165.2(1,2.8)	
				$La^{III}O_{10}$	μO 2.609(3,21)2.3		O,O 47.31(-)	
					O 66(3)		167.64(10)	
					NO_3O 2.612(3,13)			
					2.774(3)			
					MeHO 2.599(3)			
[(EtOH)Cu_3(μ-chp)$_7$La(chp).	tr	10.850(15)	79.41(8)	CuO_2N_2	O 1.948(7,6)	Cu 3.699(1)	O,O 170.8(3)	143
(NO_3)].0.5Et_2O	P-1	12.158(19)	78.02(9)		N 1.963(7)		N,N 171.2(3)	
(green)	2	21.17(3)	78.31(9)		2.003(7)		O,N 90.3(3,2.4)	
				CuO_2N_2	μO 1.969(5,2)	La 3.730(1)	O,O 81.65(22)	
					N 1.989(7,9)	O 109.3(2,6)	N,N 95.1(3)	
							O,N 94.2(2,7)	
							160.9(3,1.4)	
				$La^{III}O_8$	μO 2.584(5,14)		O,O 48.61(-)	
					O 2.387(7,29)		158.81(20)	
					NO_3O 2.612(7,7)			
				CuN_3O	N 2.016(7,13)	Cu 3.720(1)	N,N 94.5(3,2.8)	
					O 1.956(5)		168.5(3)	
							N,O 85.5(2,2.2)	
							178.4(3)	

Table 4. Continued 3

Compound	Cryst. cl. / Space gr. / Z	a, b, c [Å]	α, β, γ [°]	Chromophore	M-L [Å]	M...M [Å]	L-M-L [°]	Ref.
$[(H_2O)Cu_3(\mu\text{-chp})_7Gd(NO_3)$.	tr	10.744(1)	78.48(3)	CuO_2N_2	μO 1.958(4,2)	Gd 3.627	O,O 80.06(17)	144
(chp)].0.5MeOH	P-1	12.131(7)	78.06(6)		N 1.984(5,13)	Cu 3.691(-,4)	N,N 95.8(20)	
(dark green)	2	21.138(18)	77.30(3)				O,N 94.6(2,1.4)	
(at 150.0(2) K)							161.0(2,6)	
				CuO_2N_2	μO 1.939(4)	Gd 3.922	O,O 170.32(20)	
					N 1.985(5,14)		N,N 170.46(21)	
					H_2O 1.955(5)		O,N 90.4(2,2.1)	
				CuN_3O	N 1.987(5)	Gd 3.743	N,N 94.2(2,2.9)	
					2.012(5,6)	Cu 6.575	168.31(20)	
					O 1.965(4)		N,O 87.1(2,3.6)	
							178.51(19)	
				$Gd^{III}O_8$	μO 2.288(4,18)		O,O 50.80(15)	
					2.473(4,11)		161.03(15)	
					O 2.239(4)			
					NO_3O 2.519(5,9)			
$[(H_2O)Cu_3(\mu\text{-chp})_7Dy(NO_3)$.	tr	10.741(7)	78.02(6)	CuO_2N_2	O not given			144
(chp)].MeOH	P-1	12.136(7)	78.01(3)					
(dark green)	2	21.199(14)	77.40(4)	CuN_3O	N not given			
(at 150.0(2) K)								
				$Dy^{III}O_8$	O not given			
$[(H_2O)_{0.5}(MeOH)_{0.5}Cu_3(\mu\text{-chp})_7$	tr	10.797(7)	77.74(4)	CuO_2N_2	O not given			144
$.Er(NO_3)(chp)]$.MeOH	P-1	12.104(8)	78.10(3)					
(dark green)	2	21.286(11)	77.67(4)	CuN_3O	N not given			
(ay 150.0(2) K)								
				$Er^{III}O_8$	O not given			
2Cu + 2M :								
$[\{Cu(\mu\text{-bdd})K\}(\mu\text{-pcr})]_2.2CHCl_3$	tr	12.592(4)	90.28(2)	CuO_4	O 1.908(4,2)		O,O 90.0(2,5.6)	145
(green)	P-1	13.352(3)	93.51(2)		μO 1.917(4,0)			
	2	10.501(2)	89.13(2)					
				K^IO_8	pcrμO 2.722(5,20)		O,O not given	
					bddμO 2.854(4,33)			
					bddO 2.847(5,47)			
					2.933(5,32)			

Table 4. Continued 4

Compound	Crystal system / space group / Z	a, b, c	β (or α, β, γ)	Chromophore	M–L	M···M / angle	L–M–L	Ref.
[(18-crown-6)Rb(μ-Cl)$_2$. Cu(μ-Cl)$_2$Cu(μ-Cl)$_2$. Rb(18-crown-6)] (green)	m P2$_1$/c 2	11.100(2) 8.168(1) 22.626(5)	97.55(2)	CuCl$_4$	μCl 2.209(8,7) 2.338(7,51)	Cu 3.386(1) Rb 3.917(1) Cl 96.8(1)	Cl,Cl 90.1(1,6.9) 173.2(1,5)	146
				Rb^{I}O$_6$Cl$_2$	O 2.980(5,72) μCl 3.362(2,28)	Cl 86.7(1,7)	Cl,Cl 57.7(1)	
[Cu(μ-Et$_3$NCH$_2$CO$_2$)$_4$Ca. (η^2-NO$_3$)(η-NO$_3$)(H$_2$O)]$_2$. (NO$_3$)$_4$.5H$_2$O (blue)	m C2 2	19.444(7) 20.408(6) 15.819(4)	124.37(2)	CuO$_5$	O 1.951(7,16) μO 1.968(6) 2.301(4)	Ca 3.420(2)	O,O 79.0(2) 90.2(3,1.8) 104.3(2) 177.7(2,1.1)	147
				Ca^{II}O$_8$	O 2.445(8,27) H$_2$O 2.432(10) O$_2$NO 2.456(13) NO$_3$O 2.490(8)x2		O,O 75.1(3,4.7) 119.2(2,1.2) 152.0(2)	
[{Cu(μ-bdmap)$_2$(μ-F$_3$ac)$_2$Sr}$_2$. (μ-F$_3$ac)$_2$] (dark blue)	m P2$_1$/c 2	13.291(3) 11.238(5) 18.472(7)	91.5(2)	CuO$_3$N$_2$	μO 1.895(10,25) N 2.085(20,45) μF$_3$acO 2.43(1)	Sr 3.390(3) O 100.9(5,5)	O,O 84.7(5) N,N 108.1(6) O,N 82.8(6,4)[d]	51
				Sr^{II}O$_5$N$_2$	μO 2.48(1,1) N 2.815(2,45) μF$_3$acO 2.49(2,2) 2.60(1)		O,O 91.8(5,9.8) 169.9(5,8.0) N,N 168.5(5) O,N 62.8(4,2)[d] 88.6(5,5.5) 124.5(4,5)	
[(acac)(moeH)Ba(μ$_3$-moe)$_2$. Cu(acac)]$_2$ (dark blue)	tr P-1 1	10.797(3) 11.269(1) 12.109(1)	106.18(1) 100.93(2) 102.98(2)	CuO$_5$	μ$_3$O 1.93(3,2) acacO 1.943(3,5) 2.590(3)	Ba 3.495(1) 3.737(1)	O,O 81.82(1) 91.0(1,2.9) 104.51(1) 171.1(1,4.6)	148
				Ba^{II}O$_9$	μ$_3$O 2.763(3,15) 2.286(3,7) acacO 2.697(3,30) 2.939(4,43) moeH 2.922(4)	Ba 3.7364(3)	O,O not given	

Table 4. Continued 5

Compound	System / Space group / Z	a, b, c	α, β, γ	Chromophore	M–L	M···M	L–M–L	Ref.
[$(hfac)_2Ba(\mu_3\text{-}tdeH)_2Cu]_2$ (lihht blue)	m $P2_1/n$ 2	12.7094(2) 9.2893(2) 25.6763(4)	100.391(1)	CuO_4S_2	μ_3O 1.940(3,4) S 2.348(2,1) O 2.532(5,10)	Ba 3.5817(6) 3.6251(6) Cu 6.082(5) O 96.88(12)	O,O 83.23(14) S,S 99.5(1) O,S 88.6(1,3)[d] 171.5(1,2)	149
				$Ba^{II}O_8$	μ_3o 2.760(3) 2.835(3,6) hfacO 2.778(4,17)	O 85.97(9)	O,O 43.14(-) 177.30(11)	
[$Cu_2(\mu_3\text{-}thd)_4(\mu\text{-}mode)_4$. $Ba_2(modeH)_2$] (purple)	tr P-1 1	10.533(3) 14.693(4) 15.920(5)	101.55(2) 107.77(2) 97.52(2)	CuO_4	O 1.926	Ba 3.612(-) 3.719(1)	not given	106
				$Ba^{II}O_6$	O 2.851	Ba 3.810(1)	not given	
[$Cu_2(\mu_3\text{-}thd)_4(\mu\text{-}mop)_4$. $Ba_2(mopH)_2$] (purple)	tr P-1 1	10.607(3) 14.422(4) 15.026(4)	72.64(2) 70.59(2) 76.45(2)	CuO_4	O 1.932	Ba 3.612(-) 3.719(1)	not given	106
				$Ba^{II}O_6$	O 2.814	Ba 3.745(1)	not given	
[$Cu_2(\mu\text{-}dmae)_2(\mu_3\text{-}dmae)_2$. $Pb_2(dmso)_2I_4$].2dmso (green)	tr P-1 2	8.305(3) 12.132(3) 12.755(3)	84.86(2) 80.79(2) 73.26(3)	CuO_2N_2	μO 1.914(8) μ_3O 1.941(7) N 2.065(10,5)	Pb 3.394(2) 3.634(2) μO not given μ_3O not given	O,O 81.6(2) N,N 105.2(4) O,N 85.7(4,1)[d] 162.2(4,4.7)[d]	150
				$Pb^{II}O_4I_2$	μO 2.343(7) μ_3O 2.473(8) 2.601(7) dmsoO 2.384(9) I 3.285(1) 3.390(1)		O,O 61.0(3) 82.0(3,2.6) 139.8(3) I,I 103.59(3) O,I 83.3(2,5.4) 115.5(2,12.2) 162.2(2,6.2)	
[{Cu(μ-mct)Pb}$_2$(μ-ac)$_2$]. $(Bph_4)_2$.2dmf (dark red)	m $P2_1/n$ 2	18.890(3) 19.968(3) 12.704(2)	91.61(1)	CuO_2N_2	μO 1.90(1,0) N 1.93(1,1)	Pb 3.577(2) O 98.8(4,4.8) Cu 3.278(3)	O,O 88.4(4) N,N 84.1(5)[d] O,N 93.7(5,5)[c]	48a
				$Pb^{II}O_5N_3$	μO 2.62(1) 2.90(1) N 2.62(1,12) μacO 2.39(2) 2.94(2,14)	Pb 4.904(1)	O,O 44.2(-) 124.9(4) N,N 68.3(4,4)[d] 88.9(4) O,N 62.7(-) 115.6(4)	

Table 4. Continued 6

{[Me_2Ga(μ-dmae)(μ-pz)Cu]$_2$. (μ-pz)$_2$} (blue)	m $P2_1/c$ 2	12.018(2) 10.390(1) 13.568(3)	100.04(6)	CuN_4O	μpzN 1.984(4,35) 2.170(3) μdmaeμO 1.975(3) μdmaeN 2.134(4)	Ga 3.397(1) O 122.3(2) Cu 3.991(1)	N,N 99.0(2,11.1) 148.6(2) N,O 80.8(1)[d] 90.0(1,3.2) 166.0(1)	151
				$Ga^{III}C_2ON$	MeC 1.946(7,6) μO 1.902(3) N 1.985(4)		C,C 121.3(4) C,O 110.1(4,1.3) C,N 110.4(4,1.3) O,N 90.0(1)	
$(NBu_4)_2[Mo_2O_5${Cu(EtO)$_2$. (μ-salphen)}$_2$] (dark red)	m $P2_1/c$ 2	20.821(4) 23.133(5) 20.056(4)	117.71(3)	CuO_4	O not given	Cu 4.110	O,O not given	152
				$Mo^{IV}O_6$	O not given	Mo 3.237(4) O 116.2(6)	O,O not given	
[Cl_3Zn(μ-Cl)Cu(μ-enp)]$_2$ (dark)	m $P2_1/n$ 2	8.965(6) 20.336(8) 10.649(5)	95.20(4)	CuO_3N_2Cl	μO 1.934(4) 2.819 O 1.913(5) N 1.929(6,12) μCl 2.727(2)	Zn not given O 96.4 Cl 110.1(1)	O,O 95.4(2,3.9) N,N 84.3(3)[d] O,N 89.7(2)[c] 93.0(2) 170.0(2,1.2) O,Cl 95.4(1,3.9) N,Cl 93.5(2,4.2)	153
				$Zn^{II}Cl_4$	Cl 2.254(2,13) μCl 2.297(2)		Cl,Cl 109.5(1,4.5)	
[(phen)$_2Cu_2$(μ-ttftch)$_2$. Cd(hfacac)$_2$] (blue)	tr P-1 1	12.238(2) 12.707(5) 11.853(3)	92.95(3) 104.45(2) 63.41(2)	CuO_2N_2	μO 1.927(3,1) phenN 1.999(3,2)	Cd 3.356(-,23) Cu 5.731(1)	O,O 91.0(1) N,N 82.0(1)[d] O,N 93.2(1,1) 172.5(1,2.0)	154
				$Cd^{II}O_6$	μO 2.266(3,2) 2.319(3,8) hfacO 2.298(3,4)	Cd 3.4927(6)	O,O 82.0(2,3.7) 98.2(2,6.9) 169.5(2,8.4)	

Table 4. Continued 7

[{Hg(μ-mp)$_2$Cu(NO$_3$)}.	tr	7.041(3)	69.13(3)	CuO$_5$	μMeO 1.905(17,2)	Hg 3.180(4)	O,O	78.2(8)	155
(μ-MeCO)]$_2$	P-1	11.343(4)	76.08(3)		μmpO 2.002(20)	O 101.8(8)		93.9(8,12.1)	
(pale blue)	1	11.427(4)	84.17(3)		2.249(22)	Cu 2.957(6)		170.2(9,1.8)	
					O$_2$NO 1.982(17)				
				Hg^{II}N$_2$	μmpN 1.975(16)		N,N	174.6(9)	
					2.050(19)				
[{(bdmmpH)Cu(μ-bdmmp).	tr	13.106(6)	97.14(4)	CuO$_3$N$_2$	μHO 2.098(8)	La 3.556(2)	O,O	87.1(4,5.1)	58
(μ-OH)La(F$_3$ac)}$_2$(μ-F$_3$ac)$_4$]	P-1	17.15(1)	108.48(4)		μO 1.963(8)	HO 101.9(3)		172.3(4)	
(dark green)	1	11.152(6)	93.85(4)		O 1.900(8)	O 107.6(4)	N,N	117.9(4)	
					N 2.20(1,10)		O,N	89.7(4)	
								90.3(4,1.1)[c]	
								100.8(4,3.7)	
								137.7(4)	
				La^{III}O$_7$N	μHO 2.471(8)		O,O	65.9(-)	
					μO 2.432(8)			149.6(3)	
					O 2.523(9,19)		O,N	70.1(3)[c]	
					2.58(1)			75.1(-)	
					N 2.90(1)			141.8(3)	
[Cu$_2$(μ-chp)$_8$La$_2$(chpH)$_2$.	tr	11.549(8)	73.71(6)	CuN$_3$O	N 2.015(8,2)	La 3.900(1,4)	N,N	91.8(3,1.8)	143
(NO$_3$)$_2$](chpH)$_2$.2MeCN	P-1	13.785(10)	66.63(6)		μO 1.991(6)	O 114.1(3)		169.6(3)	
(green)	1	14.144(12)	82.37(6)				N,O	88.7(3,2)	
								173.5(3)	
				La^{III}O$_8$N	μO 2.542(6,9)	La 4.289(1)	O,O	68.3(-)	
					2.641(6)	O 115.0(2)		150.9(2)	
					O 2.368(6)		O,N	49.7(-)	
					2.472(6,8)			145.5(2)	
					N 2.765(8)				
					NO$_3$O 2.619(7,13)				

Table 4. Continued 8

[Cu(μ$_3$-OH)(μ-salpn)Dy(hfac)$_3$]$_2$ (pale blue)	tr P-1 2	12.386(5) 13.574(3) 13.971(4)	69.38(2) 65.35(6) 69.93(5)	CuO_2N_2	μ$_3$HO 1.975(10) μO 1.939(9) N 1.954(17,25)	Dy not given μ$_3$HO 98.3(3) 114.8(4) μO 98.0(5)	O,O 80.3(4) O,N 95.0(6) 169.6(5)	156
				$Dy^{III}O_9$	μ$_3$HO 2.401(7)x2 μO 2.444(9) O 2.358(9,14) 2.440(12,32)		O,O 64.5(-) 153.6(5)	
[(MeOH)(η^2-NO$_3$)$_2$(mhpH)Dy. (μ-mhp)$_2$Cu(μ-MeO)]$_2$ (not given)	tr P-1 1	9.130(9) 12.096(9) 15.222(12)	78.86(4) 82.93(8) 76.98(6)	CuO_2N_2	μMeO 1.964(4,8) μmhpN 1.980(5,6)	Cu 2.956(2)	O,O 82.4(2) N,N 97.8(2) O,N 95.8(2,4) 152.7(2,2)	157
				$Dy^{III}O_8$	mhpHO 2.256(3) μmhpO 2.221(4,10) MeHO 2.357(1) NO$_3$O 2.474(5,45)		O,O not given	
[(MeOH)(η^2-NO$_3$)$_2$(mhpH)Yb. (μ-mhp)$_2$Cu(μ-MeO)]$_2$ (not given)	m P2$_1$/c 2	11.1807(23) 17.136(10) 14.771(6)	73.98(4)	CuO_2N_2	μMeO 1.978(8,3) μmhpN 1.986(10,4)	Cu 2.966(2)	O,O 82.3(2) N,N 100.4(4) O,N 94.8(4,0) 151.8(4,7)	157
				$Yb^{III}O_8$	mhpHO 2.185(10) μmhpO 2.193(9,3) MeHO 2.302(9) NO$_3$O 2.442(10,41)		O,O not given	
[(MeOH)(η^2-NO$_3$)$_2$(mhpH)Gd. (μ-mhp)$_2$Cu(μ-MeO)]$_2$ (not given)	tr P-1 1	9.103(4) 12.003(4) 15.297(6)	81.155(36) 82.886(28) 74.789(24)	CuO_2N_2	μMeO 1.972(4,2) μmhpN 1.968(6,8)	Cu 2.971(1)	O,O 81.4(2) N,N 97.1(2) O,N 96.2(2,1.1) 153.7(2,2)	157
				$Gd^{III}O_8$	mhpHO 2.289(5) μmhpO 2.264(5,1) MeHO 2.38595) NO$_3$O 2.483(6,54		O,O not given	

Table 4. Continued 9

[(MeOH)(η^2-NO_3)$_2$(mhpH)Nd. (μ-mhp)$_2$Cu(μ-MeO)]$_2$ (not given)	tr P-1 1	9.011(3) 11.2662(24) 16.110(3)	95.506(1) 95.103(23) 88.817(17)	CuO_2N_2	μMeO 1.959(4,1) μmhpN 1.985(4,5)	Cu 2.961(2)	O,O 81.8(1) N,N 98.5(1) O,N 93.6(2,7) 151.4(2,9)	157
				$Nd^{III}O_9$	mhpHO 2.342(4) μmhpO 2.563(4,1) MeHO 2.526(4,11) NO_3O 2.594(4,62)		O,O not given	
[(MeOH)(η^2-NO_3)$_2$(mhpH)Ce. (μ-mhp)$_2$Cu(μ-MeO)]$_2$ (not given)	tr P-1 1	9.127(5) 11.503(6) 14.855(7)	72.59(4) 86.47(3) 87.99(4)	CuO_2N_2	μMeO 1.984(5,12) μmhpN 1.984(5,12)	Cu 2.974(1)	O,O 81.1(2) N,N 97.3(2) O,N 96.4(2,3) 153.3(2,4)	157
				$Ce^{III}O_9$	mhpHO 2.393(5) μmhpO 2.378(4,9) MeHO 2.569(5,47) NO_3O 2.645(6,50)		O,O not given	
[(MeOH)(η^2-NO_3)$_2$(mhpH)La. (μ-mhp)$_2$Cu(μ-MeO)]$_2$ (not given)	tr P-1 1	9.036(5) 11.340(5) 16.062(6)	85.391(24) 85.43(5) 88.25(5)	CuO_2N_2	μMeO 1.969(7,4) μmhpN 1.985(9,20)	Cu 2.980(2)	O,O 81.6(3) N,N 97.8(4) O,N 96.1(3,4) 151.8(3,9)	157
				$La^{III}O_9$	mhpHO 2.401(10) μmhpO 2.409(8,1) MeHO 2.574(9,14) NO_3O 2.648(10,63)		O,O not given	
[{(hfac)$_2$Y(μ-moe)$_2$Cu}$_2$. (μ_3-moe)$_2$] (light blue) (at 238 K)	m P2_1/n 2	10.800(4) 20.067(7) 13.703(5)	 96.42(3)	CuO_5	μ_3O 2.008(8) 2.362(9) μO 1.916(9,7) O 2.119(11)	Y 3.317(3) 3.471(2) Cu 3.276(3) μ_3O 97.3(4,2.3) μO 109.4(4,3.3)	O,O 80.4(4)[d] 81.7(4,3.2) 95.4(4,5) 102.0(4,1.7) 169.4(4,6.5)	107
				$Y^{III}O_8$	μ_3O 2.327(9) μO_2 2.239(9,4) O 2.526(11) hfacO 2.358(10,29) 2.410(11,5)		O,O 66.2(-) 145.7(4)	

Table 4. Continued 10

Cu + 3M :

Compound		a, b, c	α, β, γ	Chromophore	Distances		Angles		Ref.
[(bpy)$_2$Cu(μ-O)Mo(O). {(μ-O)Mo(O)$_2$(csh)}$_2$] (blue green)	tr P-1 2	13.381(3) 14.355(3) 14.463(2)	77.82(2) 68.00(2) 73.72(2)	CuN_4O	bpyN μO	1.984(4,6) 2.104(4,57) 1.942(3)	O 140.6(2)	N,N 95.7(2) 175.0(2) N,O 116.2(2)	158
				$Mo^{III}O_4$	μO O	1.745(4,3) 1.770(3) 1.720(4)	O 145.2(2) 161.2(2)	O,O 109.5(2,2.8)	
				$Mo^{IV}O_5N$	μO cshO cshN	2.239(4) 1.700(4,5) 1.938(4,23) 2.287(5)		O,O 104.5(2) 160.0(2,8.0) O,N 165.2(2)	

a Where more than one chemically equivalent distance or angle is present, the mean value is tabulated. The first number in parenthesis is e.s.d., the second is a maximum deviation from the mean value.

b The chemical identity of coordinated atom/ligand is specified in these columns.

c Six - membered matallocyclic ring

d Five - membered metallocyclic ring.

e Four - membered metallocyclic ring.

Table 4 A. Summary of Cu - M and M - M distances [Å] in heterotetranuclear copper(II) compounds

Cu - M	Distances [ref] M = non transition
- Hg(II)	3.180(4) [155]
- Cd(II)	3.3557(6) [154]
- Sr(II)	3.390(3) [51]
- Pb(II)	3.394(2), 3.643(2) [150]; 3.577(2) [48a]
- Ga(III)	3.397(1) [151]
- Ca(II)	3.420(2) [147]
- Ba(II)	3.495(1), 3.737(1) [148]; 3.5817(6), 3 6251(6) [149]
	3.612(1), 3.719(1) [106]
- Rb(I)	3.917(1) [146]
	M = transition
- Y(III)	3.204(2), 3.246, 3.377(2) [107]; 3.317(2), 3.471(2) [107]
	M = lanthanoide
- La(III)	3.556(2) [58]; 3.719(1) [143]; 3.730(1) [143];
	3.900(1) [143]
- Gd(III)	3.627, 3.743, 3.922 [144]
Cu(II) - Cu(II)	2.956(2), 2.961(1), 2.966(2), 2.971(1), 2.974(1), 2.980(2)
	[157]; 2.957(6) [155]; 2.9699(7) [143]; 3.107(2),
	3.711(2) [105]; 3.276(3) [107]; 3.287(3) [48a]; 3.386(1)
	[146]; 3.688, 3.695, 6.575 [144]; 3.699(1) [143];
	3.991(1) [151]; 4.110 [152]; 5.731(1) [154]; 6.082(5) [149]
Mo(VI) - Mo(VI)	3.237(4) [152]
Cd(II) - Cd(II)	3.4927(6) [154]
Ba(II) - Ba(II)	3.7364(3) [148]; 3.745(1), 3.810(1) [106}
La(III) - La(III)	4.289(1) [143]
Pb(II) - Pb(II)	4.904(4) [48a]

Table 5. Crystallographic and structural data for heteropenta- and heterooligonuclear copper(II) compounds[a].

COMPOUND (colour)	CRYST. CL. SPACE G. Z	a[Å] b[Å] c[Å]	α[°] β[°] γ[°]	CHROMO PHORE	M - L [Å]	M - M[Å] M - L - M[°]	L - M - L [°]	REF.
[(bpy)$_2$Cu$_2$(μ-bzasp)$_2$Na$_{0.5}$]$_2$.ClO$_4$. 5H$_2$O[c] (green)	m P2$_1$ 2	18.603(2) 12.943(1) 20.945(2)	 106.65(1)	CuO$_3$N$_2$	μbzaspO[b] 1.962(8,30) 2.448(10,18) bpyN 1.987(10,13)	not given	O,O[b] 59.3(3,2)[d] 96.3(3,3.7) N,N 81.5(3,1.0)[e] O,N 93.7(3,2.2) 110.7(4,1.7) 168.5(4,1.9)	159
				CuO$_3$N$_2$	μbzaspO 1.962(10,52) 2.438(8,29) bpyN 1.994(9,12)		O,O[b] 58.9(3,6)[d] 94.4(3,2.7) N,N 81.9(4,0)[e] O,N 94.4(4,4.5) 108.6(3,8) 167.4(3,3.0)	
				Na^{I}O$_8$	μbzaspO 2.40(-,7) 2.92(-,7)		O,O not given	
[{(μ-F$_3$ac)Cu$_2$(μ$_3$-deae)$_2$(μ-pyo)$_2$}$_2$ Ba] (dark green)	m P2$_1$/c 4	15.919(6) 15.903(6) 30.31(2)	 95.16(5)	CuO$_3$N$_2$	μ$_3$O 1.90(3,2) 1.99(3,5) N 2.09(6,7) μpyoN 1.93(3,3) μF$_3$acO 2.25(4,3) 2.35(4)	Ba 3.444(7,17) Cu 2.952(9,2) O 97(1,0)	O,O 74(1,2) 99(2,9) N,N 105(1,4) O,N 92(1,13) 156(2,5)	160
				Ba^{II}O$_8$	μ$_3$O 2.84(3,3) μpyoO 2.63(3,2) 2.72(3)	O 90(1,1)	O,O 49.1-161.3(8)	

Table 5. Continued 2.

[{(μ-F_3ac)Cu_2($μ_3$-bdmap)$_2$	m	25.75(2)		CuO_3N_2	$μ_3$O	1.815(20,15)	Ba	3.447(6,13)		not given	160
(μ-pyo)$_2$}$_2$Ba].0.5thf	C2	13.63(3)	115.58(7)			2.06(2,6)	O	90(1,1)			
(blue-green)	4	23.57(5)			N	2.04(3,13)		97(1)			
					μpyoN	1.99(3,12)					
					μF_3acO	2.26(3,11)					
				$Ba^{II}O_8$	$μ_3$O	2.87(2,1)	O	87.0(8,9)	O,O	50.9-167.0(9)	
						2.97(3)					
					μpyoO	2.72(3,2)					
[{Cu(μ-thd)}$_4$($μ_3$-moe)$_2$(μ-moe)$_2$	tr	14.285(5)	77.57(3)	CuO_6	$μ_3$O	1.89(2,1)	Ba	3.525(4).	O,O	78.1(5,1.4)	161
Ba(moe)$_2$]	P-1	16.407(5)	89.60(3)	(x2)		2.24(2,21)		3.664(4)		92.9(5,7.3)	
(blue)	2	19.409(6)	81.14(3)		μO	1.97(2,3)	Cu	3.000(4)		146.0(5,8.0)	
						2.75(4,6)				175.8(5,3.6)	
					O	1.91(2,3)					
						2.55(2)					
				CuO_6	$μ_3$O	1.89(2)	Ba	3.529(4)			
				(x2)		2.45(2)		3.676(4)			
					μO	1.94(2,4)	Cu	3.002(4)			
						2.77(4,4)					
					O	1.94(2,2)					
						2.63(4,10)					
				$Ba^{II}O_{10}$	$μ_3$O	2.70(2,2)			O,O	50.3-162.4(5)	
						3.08(2)					
					μO	2.75(2,2)					
						3.01(2,1)					
					O	2.87(2,2)					

Table 5. Continued 3.

$[Cl_3Zn(\mu\text{-}Cl)Cu(\mu\text{-}Cl)(\mu\text{-}attn)Cu$ $(\mu\text{-}attn)(\mu\text{-}Cl)Cu(\mu\text{-}Cl)ZnCl_3]$ (green)	m $P2_1/c$ 2	8.248(5) 28.203(3) 10.405(1)	93.07(2)	CuN_4Cl_2	N μCl	2.01(1,3) 2.878(3,0)	Cu Cl	3.620(3) 7.239(4) 80.6(1)	N,N N,Cl	90.0(5,9) 90.0(3,2.2)	162
				$CuN_2Cl_2S_2$ (x2)	N S μCl	1.99(1,1) 2.380(4,11) 2.713(4) 2.979(5)	Zn Cl	4.716(3) 6.151(3) 126.3(2)	N,N S,S N,S N,Cl Cl,S	95.8(5) 80.8(1) 91.8(4,1) 171.8(4,4) 85.2(4,4.4) 95.5(2,5.0)	
				$Zn^{II}Cl_4$	Cl μCl	2.266(6,16) 2.296(5)			Cl,Cl	109.5(2,3.1)	
$[\{(H_2O)Cu(\mu\text{-}Clac)_4Gd(H_2O)_3$ $(\mu\text{-}Clac)\}_2Cu(Clac)_2]$ (blue)	m $P2_1/c$ 2	10.521(2) 18.024(4) 15.310(3)	99.85(3)	CuO_4	μClacO ClacO	1.954(4,0) 1.970(5,0)	Gd	4.662(1)	O,O	89.3(2)	163
				CuO_5	μClacO H_2O	1.968(5,12) 2.202(5)	Gd	3.561(1)	O,O	not given	
				$Gd^{III}O_8$	μClacO H_2O	2.358(5,39) 2.442(5,10)			O,O	not given	
$[Cu\{(\mu\text{-}Br)Tl(18\text{-}crown\text{-}6)\}_4].$ $(TlBr_4)_2$[c] (not given) (at 115 K)	c F23 4	20.674(3)		$CuBr_4$	μBr	2.328(2,0)	Tl	3.5572(6)	Br,Br	not given	164
				$Tl^{I}O_6Br$	O μBr	2.96(1) 2.99(2) 3.5430(8)			O,O	56.8-143.8(4)	
				$Tl^{III}Br_4$	Br	2.551(1,5)			Br,Br	not given	
(at 295 K)				$CuBr_4$	μBr	2.311(2,0)	Tl	3.5893(8)	Br,Br	not given	
				$Tl^{I}O_6Br$	O μBr	2.91(2) 3.11(2) 3.5629(9)			O,O	55.9-142.0(5)	
				$Tl^{III}Br_4$	Br	2.544(2,9)			Br,Br	not given	

Table 5. Continued 4.

[{(η^5-cp)Fe_2(μ-cpme)}$_2$Cu$(NO_3)_2$]	tr	7.520(1)	101.93(2)	CuN_4O_2	N	2.047(5,6)	Fe	6.042(2)	N,N	84.6(2)[c]	165
.2MeOH	P-1	11.047(2)	97.24(2)		O_2NO	2.515(6,0)		6.211(2)		95.4(2)	
(yellow)	1	16.186(5)	104.95(1)				Cu	7.520(1)	N,O	90.2(2,3.1)	
				FeC_{10}	C	2.041(9,14)			C,C	not given	
[{(H_2O)Cu(μ-wil)$_2$Na(H_2O)}$_2$Pd].	or	20.893(2)		CuO_3N_2		not given				not given	166
$9H_2O$	Pbca	17.522(1)									
(not given)	8	25.758(2)		Na^IO_5	O	2.284(9,39)	Pd	2.976(5,27)		not given	
						2.383(9,49)					
				$Pd^{II}N_4$	N	2.031(9,9)			N,N	90.0(4,2.3)	
[Cu_5(μ-alaha)$_5$Nd$(H_2O)_4$].$(NO_3)_2$.	or	13.5774(1)		CuO_2N_2	μ_3O	1.968(4)			O,O	84.8(2)	167
(OH).12H_2O.MeOH	$P2_12_12_1$	16.9867(2)			μO	1.926(5)			N,N	82.7(2)[c]	
(purple)	4	21.3471(3)			N	1.907(5)			O,N	96.2(2,5.2)	
						2.025(6)					
				$Nd^{III}O_9$	μ_3O	2.568(4)			O,O	not given	
					μO	2.527(5,33)					
					H_2O	not given					

Table 5. Continued 5.

$[Sr_2Cu_4(\mu\text{-bdmap})_6(\mu_3\text{-OH})_2$	m	16.112(9)		CuO_4N	μ_3HO	1.97(2,1)	Sr	3.457(4)	O,O	91.2(7,4.7)	51
$(F_3ac)_4(thf)_2]$	$P2_1/n$	19.634(8)	111.63(5)		μO	1.87(1)	Cu	2.865(5)		173.2(7)	
(dark blue)	2	16.85(1)			O	2.40(2)			O,N	85.3(8)[e]	
					N	2.04(2)				96(1,3)	
										168.9(9)	
				CuO_3N_2	μO	1.865(20,15)	Sr	3.456(4)	O,O	86.9(8)	
					O	2.49(2)			N,N	106(1)	
					N	2.095(30,15)			O,N	83(1,1)[e]	
										167(1,3)	
				$Sr^{II}O_6N_2$	μ_3HO	2.64(1)			O,O	63.4(5,3.0)	
					μO	2.51(2,7)				82.7(6,7.3)	
					O	2.635(20,15)				139.1(6,8.7)	
					N	2.94(3,3)			N,N	108.1(8)	
									O,N	61.7(7,0)[e]	
										79.0(8,10.7)	
										116.9(7,7.7)	
										147.5(7,1.6)	

Table 5. Continued 6.

[(hfac)$_2$PrCu$_2$(μ_6-O)(μ-tde)$_3$ (μ-tdeH)$_2$Cu$_2$Pr(hfac)$_2$] (green)	tr P-1 2	12.081(2) 14.493(3) 20.640(4)	72.69(3) 75.30(3) 76.72(3)	CuO$_4$S	μ_6O 2.034(4) μ_3O 2.412(5) μO 1.895(5,3) S 2.336(2)	Pr 3.2848(11) 3.3740(14) Cu 2.7674(13) 2.980(2)	O,O 72.5(2) 88.8(2,7.6) 172.8(2) O,S 90.3(2,8.4) 154.71(13)	149
				CuO$_3$S	μ_6O 2.039(4) μ_3O 1.914(4) μO 1.902(5) S 2.358(2)	Pr 3.352(2) Cu 2.7674(13) 2.973(2)	O,O 84.5(2,5) 168.9(2) O,S 95.3(2,6.9) 165.1(1)	
				CuO$_3$S	μ_6O 2.006(4) μO 1.904(5,1) S 2.335(2)	Pr 3.3295(14) Cu 2.8968(12) 2.973(2)	O,O 85.7(2,2) 171.2(2) O,S 94.3(2,6.6) 173.30(13)	
				CuO$_3$S$_2$	μ_6O 2.137(4) μO 1.894(5,1) S 2.531(2,42)	Pr 3.370(2) 3.396(2) Cu 2.8968(12) 2.980(2)	O,O 83.9(2,7) 167.7(2) S,S 86.3(8) O,S 94.5(2,11.7) 136.8(1,3.2)	
				Pr^{III}O$_9$	μ_6O 2.661(4,8) μO 2.436(5,42) O 2.520(5,76)		O,O 61.75- 145.8(2)	
[Cu$_2$(μ-fsaaep)$_2$(μ-NO$_3$)$_2$Pr (NO$_3$)]$_2$.0.5MeOH.0.5H$_2$O (yellow)	or P2$_1$2$_1$2$_1$ 4	18.746(4) 24.196(4) 17.053(4)		CuO$_4$N$_2$	μO 2.00(2,7) N 1.99(2,8)	Pr 3.418(3,18) Cu 4.649(4,53) 5.870(4,16)	O,O 77.5(6,5.7) 100.5(6,3.1)	168
				Pr^{III}O$_{10}$	μO 2.46(2,4) μO$_2$NO 2.60(2,3) O$_2$NO 2.57(2,6)	O 99.9(6,2.8)	O,O 49.7(7,9)[d] 60.5 147.9(5)	

Table 5. Continued 7.

Compound	Class / Space group / Z	a, b, c	α, β, γ	Chromophore	M–L	M–M / angle	L–M–L	Ref.
[(thf)Cu(μ-$O_2CC_5H_4$Fecp)$_2$]$_2$	tr	9.5837(15)	92.460(17)	CuO_5	O 1.965(7,18)	Cu 2.605(1)	O,O 89.4(3,1.2)	169
(green)	P-1	10.1178(19)	91.859(16)		thf 2.226(6)		97.5(2,4.6)	
	1	13.6314(33)	104.946(14)				169.3(3,2)	
				FeC_{10}	C 2.037(11,39)		C,C not given	
{[(dien)Co(μ_3-cnda)]$_2$Cu(H_2O)}$_2$.	tr	10.867(4)	108.64(2)	CuO_5	O 1.959(7,5)	Cu 2.614(2)	O,O 92.6(3,7.1)	170
$(ClO_4)_4.8H_2O$	P-1	13.531(3)	102.70(3)		H_2O 2.166(9)		168.3(3,4)	
(dark crimson)	1	15.710(6)	94.50(2)					
				$Co^{III}N_4O_2$	N 1.936(9,11)		N,N 86.3(4,7)[e]	
					2.015(8,5)		95.7(4,4.1)	
					O 1.899(7,7)		173.9(4,6)	
							O,O 88.1(3,2)	
							N,O 86.2(3,1.0)[e]	
							89.7(4,1.1)	
							175.3(3,5)	
[(Me)$_3Ga_2$(μ-mpz)$_3$(μ_3-O)Cu]$_2$	m	10.766(4)		CuN_4O	μN 1.999(4)	Ga 3.240(1,28)	N,N 92.3(2,6.3)	171
(purple)	$P2_1/c$	14.419(2)	115.923(6)		2.871(4)	Cu 3.383(1)	176.9(2)	
	2	15.174(1)			N 1.974(4,13)	N 86.0(1)	N,O 90.1(2,6)	
					μ_3O 1.920(3)	O 118.5(2,8)	170.7(2)	
				$Ga^{III}N_2OC$	N 1.958(5,10)	Ga 3.254(1)	N,N 102.8(2)	
					μ_3O 1.823(4)	O 123.0(2)	N,O 100.7(2,5.7)	
					MeC 1.950(7)		N,C 114.9(4,5.9)	
							O,C 120.4(4)	
				$Ga^{III}C_2ON$	MeC 1.973(8,5)		C,C 122.2(5)	
					μ_3O 1.879(3)		C,O 110.3(3,7)	
					N 1.975(5)		C,N 109.0(4,2.8)	
							O,N 91.6(2)	

Table 5. Continued 8.

[Fe{(μ-CN)Cu(tpma)}$_6$](ClO$_4$)$_8$.	m	16.673(2)		CuN_4	μCN	1.93(2,2)	Fe	4.899(4)	N,N	81.8(6,1.9)[e]	172
3H_2O	P2$_1$	24.909(3)	114.74		tpmaN	2.05(1,10)		4.929(4)		97.9(6,1.8)	
(purple)	2	18.060(3)									
				$Fe^{II}C_6$	μNC	1.88(2,3)			C,C	90.0(7,2.3)	
										178.4(2,9)	
[Cu$_4$(μ$_4$-O)$_3$(μ-pr^{i}O)$_{10}$Zr$_4$(pr^{i}O)$_8$]	tr	12.673(8)	104.85(3)	CuO_4	μ$_4$O	1.888(11,8)	Zr	3.118(7,8)	O,O	86.6(5,2.1)	173
(blue)	P-1	17.482(13)	113.52(3)		μO	1.896(12,5)		3.891(6,24)		103.7(5,3)	
(at 181 K)	1	10.877(8)	75.65(3)			1.965(11,1)	Cu	2.781(8)		162.6(5.6.8)	
								3.933(8,3)			
							μ$_4$O	92.4(5,5)			
								179.78			
							μO	94.3(5)			
				$Zr^{IV}O_6$	μ$_4$O	2.222(11,9)	Zr	3.205(4)	O,O	not given	
					μO	2.121(12,14)	μ$_4$O	92.3(4)			
						2.259(12,1)	μO	94.0(4,2)			
					O	1.931(13,11)					
[Y$_2$Cu$_8$(μ-pyo)$_{12}$(μ$_4$-O)$_2$(μ-Cl)$_2$	m	21.932(5)		CuO_3N_2	μO	1.975(20,5)	Cu	3.138(7)	O,N	164(1)	174
(NO$_3$)$_4$(H$_2$O)$_2$].2H$_2$O	C2/c	23.763(7)	93.47(2)	(x2)	N	1.925(20,15)		105(1,1)			
(dark green)	4	18.683(6)			H_2O	2.33(2)					
				CuO_3N_2	μO	1.955(20,5)			O,N	167.1(9)	
				(x2)	N	1.925(20,15)					
					O_2NO	2.35(3)					
				CuO_3NCl	μ$_4$O	1.86(2,1)	Y	3.375(6,47)	O,N	172(1,2)	
				(x4)	μO	2.08(2,0)	Cu	3.029(7)	N,Cl	100.1(8,1)	
						2.20(2,3)		4.74			
					N	1.935(20,15)	μ$_4$O	109(1)			
					μCl	2.42(1,5)	μCl	77.5(4)			
				$Y^{III}O_8$	μ$_4$O	2.28(3,2)	μ$_4$O	112(1)	O,O	45(1)[d]	
					μO	2.29(2,1)	μO	99.2(9,2)		68(1)	
					O_2NO	2.47(4,3)				176.0(8,2.0)	

Table 5. Continued 9.

$[Cu_2\{\mu-((CO)_9Co_3(\mu_3-CCO_2))\}_3$	m	21.783(5)		CuO_5		not given				not given	175
$(\mu-ac)(thf)_2].0.5thf$	C2/c	15.978(1)	94.06(2)								
(dark red)	4	37.463(8)		CoC_5		not given				not given	
$(tba)_4H_3[(PO_4)W_3Mo_7Cu_2O_{34}$	c	17.673(3)		MO_6	μO	1.900(4)			O,O	39.8-160.1(5)	176
$(H_2O)_2].2H_2O$	Im3m					2.488(15)	O	90.6(7)			
(blue)	2										
$(Me_4N)_2H_2[Cu_2(H_2O)_6UMo_{12}$	trg	12.943(2)		CuO_6	μO	2.068(6)	Mo	3.034(2)	O,O	90.0(5,5.3)	177
$O_{42}].14H_2O$	R-3					2.861(7,0)	U	5.562(2)		174.3(4)	
(orange)	3	33.618(5)			H_2O	2.861(7,19)	O	146.4(4)			
				UO_n	μO	2.498(5,25)	Mo	3.515(1,12)			
				MoO_6	μO	1.705(6,19)	Mo	3.155(1)-	O,O	69.4-164.7(2)	
						1.950(6,17)		3.860(1)			
						2.273(7,50)	O	87.9(2,1.4)			
						2.570(9,90)		106.7(2)			
						2.850(11,67)		132.2(2,1.0)			
$[(\mu_{12}-ClO_4)Cu_{12}(\mu_3-OH)_{24}$	tr	17.731(6)	72.78(4)	CuO_6	μ_3HO	1.973(6,27)	Gd	3.461(2)-	O,O	77.0(4,11.0)	178
$(\mu-pyp)_{10}(pyp)_2Gd_6(H_2O)_{16}].$	P-1	17.999(8)	63.96(4)		ClO_4,μ_3O	2.353(13,65)		3.572(3)		96.9(4,9.7)	
$(ClO_4)_{17}.16H_2O$	1	18.678(4)	76.93(6)		μpypO	2.281(6,36)	Cu	3.336(3)-		159.0(3,1.9)	
(grey blue)								3.458(3)		178.2(2,1.4)	
							O	92.0(4,2.5)			
								116.8(4,4.6)			
				$Gd^{III}O_9$	μ_3HO	2.461(6,42)	Gd	6.983(1)-	O,O	51.4-143.1(2)	
					μpypO	2.407(6,28)		7.033(1)			
					pypO	2.465(6,36)					
					H_2O	2.495(7,17)					

Table 5. Continued 10.

$[(\mu_{12}\text{-}ClO_4)Cu_{12}(\mu_3\text{-}OH)_{24}$	tr	17.751(5)	72.73(4)	CuO_6	μ_3HO	1.964(7,46)			O,O	77.0(9,2.0)	178
$(\mu\text{-pyp})_{10}(pyp)_2Sm_6(H_2O)_{16}]$.	P-1	18.001(7)	64.08(4)		ClO_4,μ_3O	2.34(3,13)	O	91.7(11,4.0)		96.0(8,8.0)	
$(ClO_4)_{17}.16H_2O$	1	18.689(5)	76.95(4)		μpypO	2.300(8,59)		115.8(4,6.6)		156.0(6,6.0)	
(blue)										178.6(3,1.0)	
				CuO_6	μ_3HO	1.967(7,30)			O,O	82.1(8,3.4)	
					ClO_4,μ_3O	2.32(3,0)				109.1(9)	
				$Sm^{III}O_9$	μ_3HO	2.473(7,65)	O	104.7(3,3.3)	O,O	51.2-144.4(3)	
					μpypO	2.422(8,30)					
					pypO	2.471(8,47)					
					H_2O	2.499(9,13)					

a Where more than one chemically equivalent distance or angle is present the mean value is tabulated. The first number in parentheses is the e.s.d.,and the second is the maximum deviation from mean value.
b The chemical identity of coordinated atom or ligand is given in the columns.
c There are two crystallographically independent molecules.
d Four - membered metallocyclic ring.
e Five - membered metallocyclic ring.

Table 5A. Summary of Cu - M and M - M distances [Å] in oligonuclear copper(II) clusters[a]

Cu(II) - M	Distances [ref] M = non transition
- Ga(III)	3.420(1,28) [171]
- Ba(II)	3.446(2,19) [160]; 3.598(2,79) [161]
- Sr(II)	3.456(4,1) [51]
- Te(I)	3.573(-,16) [164]
- Zn(II)	4.176, 6.151 [162]
	M = transition
- Mo(VI)	3.042(2) [177]
- Zr(IV)	3.118(7,8), 3.891(6,24) [173]
- Y(III)	3.375(6,47) [174]
- Fe(II)	4.914(4,15) [172]
- Fe(0)	6.130(2,88) [165]
	M = Lanthanoide
- Pr(III)	3.2848(11), 3.364(1,35) [149]; 3.418(3,18) [168]
- Gd(III)	3.516(3,56) [178]; 4.662(1) [163]
- U(0)	5.562(2) [177]
Cu(II) - Cu(II)	2.605(1) [169]; 2.614(2) [170]; 2.7674(13), 2.956(-,59) [149]; 2.781(8), 3.933(8,3) [173]; 2.865(5) [51]; 2.952(2) [160]; 3.001(1,1) [161]; 3.084(7,54), 4.74 [174]; 3.383(1) [171]; 3.397(3,61) [178]; 3.620, 7.297 [162]; 4.649(4,53), 5.870(4,10) [162]; 7.520(1) [165]
Zr(IV) - Zr(IV)	3.205(4) [173]
Mo(VI) - Mo(VI)	3.155 - 3.860(1) [177]
Ga(III) - Ga(III)	3.254(1) [171]
Gd(III) - Gd(III)	3.651(1) [163]; 6.982 - 7.033(1) [178]
Na(I) - Pd(II)	2.796(5,27) [166]
Mo(VI) - U(0)	3.315(3,12) [177]

[a] The first number in parenthesis is e.s.d., the second is maximum deviation from the mean value

Table 6. Crystallographic and structural data for heteropolynuclear copper(II) compounds[a].

COMPOUND (colour)	CRYST. CL. SPACE G. Z	a[Å] b[Å] c[Å]	α[°] β[°] γ[°]	CHROMO PHORE	M - L [Å]	M - M[Å] M - L - M[°]	L - M - L [°]	REF.
$[Cu(\mu\text{-pht})_2Li_2(H_2O)_2].2H_2O$ (deep blue)	tr P-1 1	7.823(7) 10.935(15) 6.514(10)	111.8(1) 112.8(1) 87.4(1)	CuO_4	μO[b] 1.942(3) O 1.955(3) 2.770(4)	Cu 6.51	O,O[b] 49.8(1,2.5)[c] 74.8(1,4.0) 87.4(1) 113.7(1)	179
				Li^IO_4	μO 1.971(4) O 1.935(5) H_2O 1.941(6,30)		O,O 97.9(2,2.7) 106.8(2,2.9) 121.8(2,1.4)	
$[Cu(\mu\text{-pht})_2Rb_2(H_2O)_2]$ (blue)	m C2/c 4	21.919(15) 7.827(7) 11.191(10)	 101.8(3)	CuO_4	μO 1.922(9) 2.792(8) O 1.965(9)	Cu 5.94	O,O 49.3(3,3.2)[c] 71.9(2,1.7) 89.4(3,3.2) 106.7	179
				Rb^IO_8	μO 2.959(8,38) O 3.069(9) 3.328(11,90) H_2O 2.839(9) 3.069(9)		O,O not given	
$[Cu(\mu\text{-pht})Cs_2(H_2O)_2]$ (blue)	m C2/c 4	21.634(18) 8.068(8) 11.502(13)	 97.5(3)	CuO_4 Cs^IO_8	not given		not given	179
$[Cu(\mu\text{-od})_2Li_2(H_2O)_4]$ (blue violet)	m $P2_1/c$ 2	7.062(4) 11.140(6) 9.583(5)	 107.67(3)	CuO_4	μO 1.9436(17) O 1.9183(17)	Li not given	O,O 86.42(7)[d]	180
				Li^IO_4	μO 1.956(4) O 1.962(4) H_2O 1.922(5,45)		O,O 105.7(2,3.3) 116.7(2,0)	

Table 6. Continued 2.

Compound	System / Space group / Z	a, b, c [Å]	α, β, γ [°]	Chromophore	L	M–L [Å]	M	M–M [Å]	L,L	L–M–L [°]	Ref.
$[Cu(\mu\text{-}C_7H_{12}O_5)_2Li_2(H_2O)_4]$	or	8.011(6)		CuO_4	μO	1.946(2,2)			O,O	87.2(1,2)[d]	181
(blue)	$P2_12_12_1$	12.205(10)			O	1.918(2,8)					
	4	22.336(13)									
				Li^IO_4	μO	1.962(6,30)	Li	2.529(8)	O,O	94.1(3,4.7)	
					O	1.964(6)				119.3(3,3.0)	
					H_2O	1.975(5)					
				Li^IO_5	μO	2.027(6,50)			O,O	91.7(2,5.1)	
					H_2O	1.979(6)				102.9(2,5)	
						2.772(9)				141.1(3)	
$[Cu(\mu\text{-eth})Li_2(H_2O)_4].2H_2O$	tr	4.849(2)	92.02(3)	CuO_4	μO	1.931(2,10)			O,O	87.21(6)[d]	182
(violet)	P-1	7.907(4)	99.23(2)								
	1	8.643(4)	103.35(2)	Li^IO_4	μO	1.919(4)					
					H_2O	not given					
$[Cu(\mu_3\text{-eryan})_2Li_3].5H_2O$	or	8.2221(13)		CuO_4	μ_3O	1.929(2,5)			O,O	87.3(1,2)[d]	182
(blue)	$P2_12_12_1$	12.112(2)									
	4	22.855(3)		Li^IO_4	μ_3O	1.921(4,7)			O,O	not given	
				Li^IO_4	μ_3O	1.938(5)					
						2.186(5)					
					O	1.976(6)					
				Li^IO_4	O	1.889(5,0)					
$[Li_{11}Cu_4(\beta\text{-cdH}_{11.5})_2].40H_2O$	or	19.182(3)		CuO_4	O	not given					183
(blue)	$P2_12_12_1$	27.062(4)									
	4	35.754(9)		Li^IO_3	O	not given					
$[Li_{11}Cu_4(\beta\text{-cdH}_{11.5})_2].25H_2O$	or	16.134(12)		CuO_4	O	not given					183
(blue)	$P2_12_12_1$	27.551(20)									
	4	33.235(33)		Li^IO_3	O	not given					

Table 6. Continued 3.

Compound (colour)	Cryst. cl., Space group, Z	a, b, c	β	Chromophore	L	Cu–L	M	M–M	L,L	Angle	Ref.
$[(H_2O)Na_2Cu(\mu\text{-pht})(pht)].H_2O$ (blue)	m $P2_1/c$ 2	11.070(10) 6.650(8) 12.143(11)	101.6(1)	CuO_4	O	1.953(4,17) 2.755(5) 3.068(4)	Cu	6.92	O,O	46.4(1)[c] 72.1(2,12.0) 100.1(2,12.0) 133.6(1)	184
				Na^IO_6	O H_2O	2.438(5,8) 2.517(5) 2.781(5) 2.286(5)			O,O	65.3 161.4(2)	
$[Cu(\mu\text{-pht})_2K_2].2H_2O$ (blue)	m $C2/c$ 4	12.065(15) 21.327(16) 7.825(9)	113.9(1)	CuO_4	μO	1.965(5,35) 2.533(5) 3.062(5)	Cu	5.74	O,O	51.6(1,5.0)[c] 76.1(2,2.0) 90.5(2,4.4) 114.9(2,4.8) 134.7(2,6.3) 175.0(1)	184
				Na^IO_6	μO	2.652(6,36) 2.770(7,36) 3.146(6,50)			O,O	not given	
$[K_2Cu_4(\mu\text{-EtSiO}_2)_6][K_2(EtSiO_2)_6]$. 4BuOH[e] (not given)	m $P2_1/c$ 4	22.97(1) 21.56(1) 16.239(9)	103.29(4)	CuO_4	μO O	1.936(8,34) 1.944(7,18)			O,O	78.8(3,3) 93.9(3,2.8) 173.1(3,2)	185
				K^IO_6	O	not given					
				CuO_4	μO O	1.933(8,25) 1.940(7,17)			O,O	78.5(3,3) 93.9(3,2.2) 173.1(3,2)	

Table 6. Continued 4.

[(ac)$_2$Cu$_3$(μ-ac)$_4$K$_2$(μ-ac)$_2$] (light green)	m C2/c 4	17.956(4) 14.312(3) 10.443(2)	 98.85(3)	CuO$_4$	μacO	1.960(4,31) 2.464(3)			O,O	57.9(1) 93.3(2,4.2) 141.8(2,10.0)	186
				CuO$_5$ (x2)	μacO acO	1.981(4,3) 2.148(3) 1.952(4,0)	Cu	2.633(1)	O,O	89.8(2,1.5) 101.7(1,1.3) 167.8(1,0)	
				K^{I}O$_8$	μacO	2.732(4,16) 2.826(3,84) 3.038(3)			O,O	not given	
[Cu(μ-pht)$_2$Mg(H$_2$O)$_2$] (not given)	or P2$_1$2$_1$2$_1$ 2	6.551(7) 20.589(10) 6.832(8)		CuO$_4$	O	1.940(4,7) 2.851(5)			O,O	91.3(2,8.5) 173.9(2)	187
				Mg^{II}O$_6$	O H$_2$O	2.031(5,0) 2.062(6,8)			O,O	90.5(4,6.7) 172.9(2,1.4)	
[Cu(μ-pht)$_2$Sr(H$_2$O)$_3$] (not given)	m C2/c 4	22.144(18) 9.122(9) 9.083(8)	 103.6(1)	CuO$_6$	O	1.967(6,4) 2.402(5,0)			O,O	90.1(3,6.9)	187
				Sr^{II}O$_6$	O H$_2$O	2.678(5,95) 2.569(8,19)			O,O	not given	
[Cu(μ$_3$-ox)$_2$Ba(H$_2$O)$_6$] (brown)	tr P-1 2	6.5405(9) 9.202(3) 10.939(1)	85.46(2) 79.22(1) 80.45(2)	CuO$_4$	μ$_3$oxO	2.759(8,2) 2.864(7,36)	O	141.9(2)	O,O	85.3(2)[d] 169.9(3)	188
				Ba^{II}O$_{11}$	μ$_3$oxO H$_2$O	2.964(6,52) 3.128(7,30) 2.759(8,2) 2.864(7,36)			O,O	42.5(2,1)[d] 62.4(2)- 151.8(2)	

Table 6. Continued 5.

$[(H_2O)_8Cu_3(\mu\text{-cdta})_2Ni_2](NO_3)_2$.	m	21.404(6)		CuO_4	μcdtaO	1.921(4)			O,O	90.4(2)	189
$7H_2O$	I2/a	7.742(2)	90.99(2)		H_2O	1.951(5)					
(not given)	4	30.712(5)									
				CuO_5	μcdtaO	1.955(4,2)			O,O	91.8(1,5.4)	
					H_2O	1.975(5,22)				171.8(1,1)	
						2.263(5)					
				$Ni^{II}O_4N_2$	O	2.005(4,99)			O,O	90.8(1,6.7)	
						2.268(4,2)				111.1(1)	
					N	2.035(5,27)				177.0(1)	
									N,N	87.2(1)	
									O,N	86.5(1,8.8)	
										164.1(1,2.8)	
$[Cu(\mu\text{-phb})_4Na_2]$	tg	16.790(7)		CuN_4	N	1.96(1,0)			N,N	not given	190
(violet)	P4/n cc	16.790(7)									
	4	20.56(1)		Na^IO_8	O	not given			O,O	not given	
					H_2O	not given					
$[(en)_2Cu(\mu\text{-ox})_2Pt]$	tr	6.978(4)	73.14(5)	CuN_4	N^b	2.02(2,0)			N,N^b	$86.0(8)^d$	191
(red-brown)	P-1	9.418(6)	100.59(4)								
	2	12.351(8)	111.73(4)	CuN_4O_2	N	2.00(2,1)			N,N	$86.4(7)^d$	
					O	2.55(1,0)					
				$Pt^{II}O_4$	O	2.02(1,0)	Pt	3.554(2)	O,O	$83.6(4)^d$	
								3.855(2)			
$[Cu(\mu\text{-}S_2C_2O_2)_2Mn(H_2O)_3]$.	m	11.692(2)		CuS_4	S	2.274(7,14)	Cu	3.681(1)	S,S	88.4(2,7)	192
$4.5H_2O$	$P2_1/c$	20.665(5)	103.84(2)							$91.7(2,2)^d$	
(dark brown)	4	7.360(2)								177.8(3,3)	
				$Mn^{II}O_7$	O	2.31(1,4)			O,O	$69.2(5,3)^d$	
					H_2O	2.16(1,2)				74.0(5,2.4)	
										90.0(5,5.0)	
										178.4(5)	

Table 6. Continued 6.

[Cu$_4$(μ-dmap)$_2$(μ-ac)$_2$(μ$_3$-OH)$_2$ Hg$_3$(μ-ac)$_2$Cl$_6$] (dark blue)	m P2$_1$/n 2	12.524(8) 12.358(7) 19.563(8)	 94.24(5)	CuO$_3$N	μ$_3$HO 1.93(1,2) μacO 1.92(2,1) μdmapO 1.93(2,1) μdmapN 2.02(2,1)	Hg 3.669(4,43) Cu 3.159- 4.832(7) μ$_3$O 113.3(7,2.8)	O,O 93.2(7,1.2) 172.8(7,2) O,N 86.7(10,3.5) 169.5(9,3)	193
				Hg^{II}O$_2$Cl$_2$	μ$_3$HO 2.44(2,1) Cl 2.327(9,29)		O,O 91.6(5) Cl,Cl 154.2(4)	
				Hg^{II}O$_2$Cl$_2$	μacO 2.76(2,0) Cl 2.300(7,0)		O,Cl 98.5(5,11.8) Cl,Cl 180.0	
[(Cu(μ-salen))$_2$NaClO$_4$].p-xylene (purple)	m C2/c 4	24.44(1) 11.283(6) 14.766(4)	 101.22(3)	CuO$_2$N$_2$	μsalenO 1.90(-,0) μsalenN 1.94(-,2)	Cu 4.52	N,N 86.5(2)[d] N,O 93.3(-,6)[f] 176.5(3,4) O,O 87.2(1)	194
				Na^{I}O$_4$	μsalenO 2.36(-,1) O$_2$ClO$_2$ 2.55(-,0)		O,O 58.5(-,8.5) 90.7(-,6.7) 111.0(-,2.0) 159.0(-,3.0)	
[(H$_2$O)$_2$Cu(μ-ic)$_4$Na$_2$(H$_2$O)$_2$]. 2H$_2$O (lilac)	tr P-1 1	10.459(1) 10.434(3) 6.762(4)	71.73(4) 83.67(4) 62.82(2)	CuO$_2$N$_2$	H$_2$O 1.937(6,0) N 1.941(9,0)		O,N 90.4(3)	195
				Na^{I}O$_6$	μO 2.389(7,59) 2.532(8,61) O 2.398(7) H$_2$O 2.470(5)	Na 3.860(5,83) O 103.3(3,3.3)	O,O 73.4-104.9(3) 165.6(3,6.0)	
[Cu(μ-ahmp)$_2$K(H$_2$O)$_2$]Br (brown)	or Pcca 4	18.482(2) 9.805(1) 8.776(3)		CuO$_2$N$_2$	O 1.915(4,0) N 1.988(4,0)		O,N 85.3(3)[d] 94.7(2)	196
				K^{I}O$_8$	O 2.848(8,97) H$_2$O 2.988(4,0)		O,O not given	

Table 6. Continued 7.

Compound				Chromophore							Ref.
[Cu(μ-ahmp)$_2$K(H$_2$O)$_2$]F.H$_2$O	or	18.814(2)		CuO$_2$N$_2$	O	1.901(3,0)			O,N	85.6(1)[d]	196
(light brown)	Pcca	9.623(1)			N	1.980(4)				94.4(1)	
	4	8.678(2)									
				K^{I}O$_8$	O	2.820(4,53)			O,O	not given	
					H$_2$O	2.965(4,0)					
[(en)Cu(μ-V$_2$O$_6$)]	m	5.772(2)		CuO$_2$N$_2$	μO	1.969(3,9)				not given	197
(green)	P2$_1$/n	8.143(3)	90.43(2)		enN	1.985(4,4)					
	4	17.483(2)									
				V^{V}O$_4$	μO	1.733(3,63)				not given	
					O	1.615(3,6)					
						1.785(3,6)					
[Cu(μ-obzp)Mn(H$_2$O)$_3$].H$_2$O	m	7.814(1)		CuO$_2$N$_2$	O	1.932(1,45)			O,O	86.41(4)	198
(red-brown)	P2$_1$/n	10.3375(7)	98.49(1)		N	1.928(1,33)			N,N	85.85(4)[d]	
	4	19.981(1)							O,N	93.8(1,6)[f]	
										178.5(1,3)	
				Mn^{II}O$_6$	O	2.173(1,34)			O,O	75.35(4)[d]	
						2.177(1)				89.7(1,8.7)	
					H$_2$O	2.151(1,28)				111.87(4)	
						2.214(1)				165.8(1,10.7)	
[Cu(μ-obbz)Mn(H$_2$O)$_3$].dmf	or	10.3006(6)		CuO$_2$N$_2$	O	1.913(3,5)	Mn	4.816	O,O	87.4(1)	199
(not given)	Pbca	22.375(2)			N	1.929(3,5)		5.506	N,N	87.0(1)[d]	
	8	19.639(1)					Cu	5.279	O,N	93.0(1,2)[f]	
										175.1(1,3.2)	
				Mn^{II}O$_6$	O	2.162(2,17)			O,O	72.6(1)[d]	
						2.239(2)				89.4(2,7.5)	
					H$_2$O	2.191(4,60)				114.6(1)	
										158.6(1,4.8)	
										175.5(2)	

Table 6. Continued 8.

Compound	Cryst. cl.	Unit cell (Å)	Angles (°)	Chromophore		M–L (Å)		M–M (Å)		L–M–L (°)	Ref.
[(Cu(μ-mesalen))$_3$Fe(acac)](PF$_6$)$_2$	hx	14.471(3)		CuO$_2$N$_2$	μO	1.905(8)			N,N	86.3(8)	200
(black)	P6$_1$22				N	1.96(1)			O,N	92.4(4)	
	6	50.372(2)								165.7(4)	
				CuO$_3$N$_2$	μO	1.932(8,9)	Fe	3.033(2)	O,O	80.3(3)	
				(x2)		2.357(8)				101.1(3,7.4)	
					N	1.945(10,5)			N,N	88.5(4)[c]	
									O,N	92.4(4,1)[f]	
										99.2(4,1.0)	
										159.1(4,6.5)	
				Fe^{III}O$_6$	μO	2.037(8,24)			O,O	75.4(3)	
					acacO	1.967(8,0)				87.7(5)[f]	
										93.4(3,4.6)	
										165.4(4,7.3)	
[Cu(μ-opba)Mn(H$_2$O)$_2$].dmso	tr	7.300(7)	68.43(3)	CuO$_2$N$_2$	O	1.950(5,2)	Mn	5.387(1)	O,O	105.9(2)	201
(violet)	P-1	10.734(3)	84.92(4)		N	1.901(4,1)			N,N	83.5(2)[d]	
	2	12.585(6)	73.97(3)						O,N	85.4(2,3)[d]	
										168.4(2,3)	
				Mn^{II}O$_6$	O	2.172(5,1)	Mn	5.030(1)	O,O	77.2(2,1)[d]	
						2.209(5,7)				90.8(2,3.2)	
					H$_2$O	2.133(5,9)				110.2(1)	
										172.4(2,1)	
(rad)$_2$Mn$_2$(Cu(μ-opba))$_3$.dmso.	m	25.379(8)		CuO$_2$N$_2$	O	not given					202
2H$_2$O	Cc	25.146(8)	131.52(4)		N	not given					
(not given)	4	18.845(4)									
				Mn^{II}O$_x$N$_x$	O	not given					
					N	not given					

Table 6. Continued 9.

Compound				Chromophore							
$[Cu_3(\mu\text{-opba})_3Tm_2(H_2O)_2].10dmf.$	or	9.864(2)		CuO_2N_2	O	1.88(1,12)	Tm	5.657(16,7)	O,O	85.3(3.3,5.2)	203
$4H_2O$	$P2_12_12_1$	20.855(2)				2.17(1,12)			N,N	76.6(3.9)[d]	
(blue)	4	39.101(9)			N	1.77(6)			O,N	86.7(3.5,3.5)	
						1.96(6,2)				163.7(3.6,7.3)	
				$Tm^{III}O_7$	O	2.189(91,16)			O,O	70.4(3.1)[d]	
						2.297(80,1)					
					H_2O	not given					
$[(H_2O)_2Cu(\mu\text{-citH})(\mu\text{-cit})Sb].$	tr	9.320(7)	104.31(4)	CuO_5	μcitHO	1.926(7,2)			O,O	85.6(3,5.6)	204
$2.5H_2O$	P-1	10.468(8)	90.01(5)		μcitO	2.015(2)				94.7(4,1.3)	
(blue green)	2	12.244(8)	110.90(4)		H_2O	2.092(6,4)				119.8(3,6.8)	
										170.9(2)	
				$Sb^{III}O_4$	μcitHO	1.992(5,18)			O,O	79.1(2,3.5)	
					μcitO	2.174(6)				97.5(2)	
						2.230(6)				146.6(2)	
$[Cu_2(\mu\text{-pht})_2(\mu_3\text{-OH})Ag(H_2O)_2].$	tr	11.388(11)	100.3(2)	CuO_5	μ_3HO	1.911(11)			O,O	90.0(4,4.5)	205
$3H_2O$	P-1	11.772(12)	80.2(1)		O	1.923(11)				135.1(4,5.3)	
(light blue)	2	7.875(9)	102.2(1)			2.128(11,105)				173.5(4)	
				CuO_5	μ_3HO	1.952(9,3)			O,O	90.0(4,6.7)	
					O	1.926(10,2)				170.5(4,4.9)	
						2.352(13)					
				$Ag^{I}O_3C_2$	O	2.264(12)			O,O	87.8(8,5)	
					H_2O	2.432(26,104)				106.1(6)	
					C	2.529(17,43)			O,C	93.4-141.7(5)	

Table 6. Continued 10.

[(Cu(μ-pba))$_3$Gd$_2$(μ-ox)]. [Cu(H$_2$O)$_5$].20H$_2$O (blue)	m C2/m 4	21.186(7) 21.098(2) 15.079(1)	92.06(2)	CuO$_3$N$_2$	μpbaO μpbaN H$_2$O	1.964(6,8) 1.923(6,2) 2.43(1)	Gd Cu	5.2799(4)- 7.2361(11) 3.6847(25) 5.686(1,101)	O,O N,N O,N	90.9(4,3.5) 96.8(3)[f] 84.5(3,1)[d] 96.2(3,6) 173.8(3,4)	206
				CuO$_5$	μpbaO H$_2$O	1.936(9,6) 2.82(3)			O,O	83.7(3,1.0)[d] 91.6(3,4.7)	
				Gd^{III}O$_9$	μpbaO oxO H$_2$O	2.454(6,22) 2.390(6,2) 2.454(6)	Gd	6.336(1)	O,O	68.6(2,3.2) 77.0(2,1.5) 90.9(2,5.3) 123.6(2,7.8) 138.1(2,6.8)	
[(SCN)$_2$Cu(μ-hoat)(μ-NCS)$_2$ Na$_2$S$_2$] (blue)	m P2$_1$/a 2	8.492(3) 17.740(4) 13.016(3)	94.54(3)	CuN$_5$	SCN μSCN μhoatN	1.97(2) 2.24(2) 2.03(2,3)	Na Cu	5.698(8,12) 6.367(4) 7.206(4)	N,N	84.9(7,1)[d] 96.3(8,4.5) 164.8(8)	207
				Na^{I}S$_3$	μNCS S	2.425(2,5) 2.41(1)			S,S	120.0(5,6.3)	
[(en)$_2$Cu(μ-NCS)$_2$Ag(SCN)$_2$] (blue-violet)	or P2$_1$2$_1$2$_1$ 4	7.126(2) 12.587(5) 17.585(5)		CuN$_5$	enN μSCN	2.007(3,17) 2.388(10)			N,N	84.0(4)[d] 93.3(4,2.8)	208
				Ag^{I}S$_4$	μNCS NCS	2.627(3,66) 2.546(3,14)			S,S	103.1(1,3.4) 112.5(1,3.3)	
[((diene)Cu)$_2$(μ-NC)$_2$Fe(CN)$_4$]. 6H$_2$O (dark green)	m C2/c 8	27.438(3) 7.778(1) 14.869(2)	100.9	CuN$_5$	dienN μCN	2.03(1,0) 1.97(1) 2.21(1)			N,N	83.6(5,4)[d] 96.6(5,4.9)	209
				Fe^{II}C$_6$	NC μNC	1.91(1,0) 1.895(10,5)			C,C	87.9(5,2.8)	

Table 6. Continued 11.

[((diene)Cu)$_2$(μ-NC)$_4$Fe(CN)$_2$]$^{+}$.	m	20.332(6)		CuN$_5$	dienN	2.009(10,19)	N,N	83.6(6,7)[d]	210
[(H$_2$O)(dien)Cu(μ-NC)$_2$Fe(CN)$_6$]$^{-}$.	P2$_1$/c	14.162(5)	96.29(3)		μCN	1.985(10,25)		96.5(6,10.5)	
4H$_2$O	4	14.932(5)				2.32(1,5)		163.4(6,7.5)	
(dark green)									
				Fe^{III}C$_6$	NC	1.955(10,15)	C,C	90.0(8,4.1)	
					μNC	1.96(2,3)		176.4(6,8)	
				CuN$_4$O	dienN	2.01(2,2)	N,N	84.2(7,1.5)[d]	
					μCN	1.99(1)		95.8(7,1.5)	
					H$_2$O	2.41(1)		171.5(7,6.8)	
							N,O	92.6(7,8.0)	
				Fe^{III}C$_6$	NC	1.938(20,23)	C,C	90.0(8,5.6)	
					μNC	1.93(2)		178.3(8,1.6)	
[Cu(μ-ntac)Li(H$_2$O)$_3$]	or	11.00(2)		CuO$_4$N	O	1.96(-,4)	O,O	93.8(5,6.9)	211
(grey)	P2$_1$2$_1$2$_1$	10.66(2)				2.26	O,N	79.5(5)[d]	
	4	9.77(2)			N	2.09		84.8(5,3)	
				Li^{I}O$_4$	O	2.06(6)	O,O	98(3)	
					H$_2$O	1.95(6,9)		111(3,13)	

Table 6. Continued 12.

$[Cu_2(\mu_3\text{-saladhp})_2(\mu_3\text{-OMe})_3$	m	12.116(1)		CuO_4N	μ_3O	1.863(3)	Mo	3.452(5,11)	O,O	77.7(1,2.8)	212
$(MoO_2)_2]$.2MeCN	$P2_1/c$	14.112(1)	116.05(1)			1.995(2,5)		4.722(5)		99.9(1,2.4)	
(green)	4	22.031(2)			N	1.917(3)	Cu	2.951(5)		178.7(1)	
					μ_3MeO	2.411(2)			O,N	84.4(1)[d]	
										96.9(1)[f]	
										106.0(1)	
										165.0(1)	
				CuO_5N	μ_3O	1.874(2)	Mo	3.532(5,3)	O,O	75.8(1,4.9)	
						2.009(2,4)		4.289(5)		100.0(1,1.2)	
					N	1.915(3)				177.1(1)	
					μ_3MeO	2.510(2)			O,N	80.5(1)[d]	
					μ_3O	2.640(22)				97.0(1)[f]	
										103.8(1)	
										163.9(1)	
				$Mo^{VI}O_6$	O	1.695(3,10)	Mo	3.528(4)	O,O	73.8(1,4.5)	
					μ_3O	2.292(2)				82.3(1)[f]	
						1.892(2)				94.1(1,9.9)	
					μ_3MeO	2.055(2,10)				150.3(1)	
						2.267(3,18)				163.1(1,3.3)	
$[(bpy)_2Cu(\mu\text{-}V_2O_6)]$	m	10.478(2)		CuN_4O	bpyN	1.979(4,10)				not given	197
(blue)	$P2_1/a$	13.779(2)	110.39(1)			2.111(4,29)					
	4	14.513(2)			μO	1.937(3)					
				V^VO_4	μO	1.678(3)				not given	
						1.777(3,3)					
					O	1.615(3)					
				V^VO_4	μO	1.834(3,4)				not given	
					O	1.621(3,8)					

Table 6. Continued 13.

$[Cu_2(\mu\text{-oxpno})(\mu\text{-NC})_4Fe(CN)$	or	11.413(4)		CuN_4O	N	1.98(2,3)	Fe	5.04	N,N	95.1(6)[f]	213
$(NO)]$	Pnma	22.242(5)			O	2.01(1)	Cu	5.28		95.1(6,7.6)	
(green)	4	7.736(2)			μCN	1.99(1)				157.1(5)	
						2.25(1)			N,O	82.4(5)[d]	
										94.8(5,3.1)	
										171.8(6)	
				$Fe^{III}C_5N$	ON	1.69(2)			C,C	87.0(9,1.5)	
					NC	1.93(3)				172.3(8)	
					μNC	1.93(2,2)			C,N	93.8(8,3)	
										179(1)	
$[Cu(\mu\text{-tsgly})_2K_2]$	tr	9.662(2)	80.36(1)	CuO_3N_2	O	1.939(4,12)			O,O	90.6(3)	214
(blue)	P-1	10.200(1)	88.24(1)			2.717(5)				171.9(4)	
	2	12.921(2)	62.94(1)		N	1.986(7,17)			N,N	164.9(4)	
									O,N	83.7(3,2)[d]	
				K^IO_6	O	2.675(5,5)			O,O	not given	
						2.749(7,35)					
						2.858(6,38)					
$[(H_2O)Cu(\mu\text{-pba})Mn(H_2O)_2]\cdot$	or	12.945(1)		CuO_3N_2	O	1.990(1,0)	Mn	5.412	O,O	94.0(1,1.2)	215
$2H_2O$	Pnma	21.250(4)			N	1.933(1,0)			N,N	95.66(8)[f]	
(light blue)	4	5.2105(8)			H_2O	2.281(3)			O,N	83.90(5)[d]	
										99.25(6)	
										165.42(6)	
				$Mn^{II}O_6$	O	2.167(1,2)	Mn	6.977	O,O	77.68(5)[d]	
					H_2O	2.197(2,0)				92.5(1,9.8)	
										180.0(1,0)	
$[(H_2O)Cu(\mu\text{-pbaOH})Mn(H_2O)_2]$	or	12.351(7)		CuO_3N_2	O	not given	Mn	5.751		not given	216
(blue)	$P2_12_12_1$	21.156(11)			N	not given	Cu	6.545			
	4	5.073(10)									
				$Mn^{II}O_6$	O	not given	Mn	6.997		not given	
					H_2O	not given					

Table 6. Continued 14.

Compound (colour)	Cryst. cl. / Space group / Z	a / b / c	α / β / γ	Chromophore	L	M–L	M	M···M	L–M–L	Angle	Ref.
[(H$_2$O)Cu(μ-pbaOH)Mn(H$_2$O)$_2$]. 2H$_2$O (not given)	m P2$_1$/c 4	21.212(4) 5.033(1) 14.246(2)	 100.72(2)	CuO$_3$N$_2$	O N H$_2$O	1.976(2,9) 1.933(3,7) 2.326(3)	Mn Cu	5.396(-,6) 5.033(1)	O,O N,N O,N	92.0(1,4.4) 96.9(1)[f] 84.2(1,1)[d] 100.4(1,3.1) 166.9(1,1.6)	217
				Mn^{II}O$_6$	O H$_2$O	2.167(2,24) 2.209(3,2)	Mn	5.033(1)	O,O	77.5(1,6)[d] 89.8(1,2.0)	
[(dmso)Cu(μ-opba)Mn(dmso)$_2$] (blue)	m P2$_1$/n 4	16.969(7) 8.367(1) 18.258(7)	 not given	CuO$_3$N$_2$	O N dmsoO	1.975(3,10) 1.910(3,2) 2.305(3)	Mn Cu	5.398(1,4) 7.505(1)	O,O N,N O,N	99.5(1,6.8) 82.4(1)[d] 84.4(1,1)[d] 95.8(1,4) 163.0(1,9)	218
				Mn^{II}O$_6$	O dmsoO	2.196(3,32) 2.139(3,8)	Mn	7.209(2)	O,O	76.0(1,1)[d] 93.3(1,11.6) 164.4(1,1.5)	
[Cu(μ-opb)Mn(H$_2$O)$_3$].H$_2$O (dark blue)	tr P-1 2	8.538(2) 9.248(8) 10.687(3)	67.73(5) 63.28(2) 74.08(5)	CuO$_3$N$_2$	O N	1.955(2,8) 2.625(1) 1.933(2,6)	Mn Cu	5.452(2) 6.066(2) 5.118(2)	O,O N,N O,N	83.39(20) 90.5(1,3) 83.97(8)[d] 92.3(1,2)[f] 100.4(1,7.6) 171.7(1,3.5)	219
				Mn^{II}O$_6$	O H$_2$O	2.085(1) 2.178(2,17) 2.103(2) 2.249(2,5)	Mn	4.819(2)	O,O	76.44(6)[d] 91.4(1,11.8) 166.5(1,6.9)	

Table 6. Continued 15.

$[(Cu(\mu\text{-saltn}))_2Ce(NO_3)_3].MeNO_2$ (not given)	or Fdd2 16	19.479(2) 26.980(2) 30.698(2)		CuO_3N_2	μO N	1.935(10,5) 2.56(2) 1.96(2,20)	Ce	3.325(2)	O,O N,N O,N	81.1(4,2.7) 97.6(7) 93.0(6,3.5) 170.2(6,4)	220
				CuO_4N_2	μO O N	2.00(1,10) 2.28(2) 2.54(2) 1.96(7,30)	Ce	3.288(2)	O,O N,N O,N	84.4(7,5.8) 167.1(9) 98.3(6) 92.8(8,5.5) 171.4(6,2)	
				$Ce^{III}O_{10}$	μO	2.50(1,8) 2.68(3,0)			O,O	not given	
$[(H_2O)_3Cu_3(\mu\text{-pba})_3Sm].20H_2O$ (dark-blue)	or Pcc2 4	16.076(5) 16.079(3) 21.256(10)		CuO_3N_2	O N H_2O	not given not given not given	Sm	5.672-5.742		not given	221
				$Sm^{III}O_8$	O	not given	Sm	11.357		not given	
$[(H_2O)_3Cu_3(\mu\text{-pba})_3Dy_2(H_2O)_2]$ (not given)	m $P2_1/n$ 4	18.637(6) 16.025(2) 19.685(6)	112.37(3)	CuO_3N_2	O N H_2O	1.98(1,2) 1.93(1,5) 2.44(2,9)	Dy	5.658(2,57)	O,O N,N O,N	90.3(7,7.2) 96.4(7,1.2)[f] 84.0(6,1.0) 98.3(6,4.9) 173.7(8,6.8)	222
				$Dy^{III}O_8$	O H_2O	2.36(1,6) 2.37(2,2)	Dy	11.939(1)	O,O	68.9-147.9(4)	

Table 6. Continued 16.

[(2-pyc)Cu(μ-2-pyc)(μ-1,3-N_3)	tr	7.499(3)	111.44(3)	CuN_3O_2	N	1.965(3,4)	N,N	92.2(1,3.7)	223
$Na(H_2O)_2$]	P-1	10.385(4)	106.30(3)		O	1.951(3)		175.8(2)	
(blue)	2	11.430(4)	94.62(3)		μO	1.944(3)	O,O	171.9(1)	
					μN_3	2.391(4)	N,O	83.4(1,0)[d]	
								95.2(1,1.7)	
				$Na^{I}O_5N$	O	2.355(3)	O,O	49.0(1)[c]	
						2.555(4)		78.5(1,4.8)	
					μO	2.750(3)		111.3(1,8.9)	
					μN_3	2.388(4)		153.9(1,6.8)	
							O,N	98.9(2,5.6)	
								139.8(2)	
[Cu(μ-sb)Mn$(CO)_3$][e]	m	9.712(2)		CuN_3O_2	N	1.84(1)	N,N	80.9(5)[d]	224
(not given)	$P2_1$	22.286(4)	94.44(2)			2.03(1,2)		104.8(5)	
	4	9.5017(7)			O	1.99(1)		158.4(5)	
						2.40(1)	O,O	93.1(4)	
							N,O	83.7(5)[d]	
								95.3(5,14.0)	
				MnC_8	η^5C	1.77*	C,C	89.7(10,1.3)	
					OC	not given			
[ClCu(μ-tacac)LiCl$(H_2O)_2$]	m	11.674(4)		CuN_3OCl	N	2.029(4,27)	N,N	84.3(2,1.2)[d]	225
(blue)	$P2_1/c$	7.543(2)	106.59(8)			2.204(4)	N,O	83.1(1)	
	4	17.560(5)			O	1.959(3)		106.2(1)	
					Cl	2.330(1)		165.0(1)	
							N,Cl	96.7(1)	
								108.7(1)	
								166.9(1)	
							O,Cl	91.5(1)	
				$Li^{I}O_3Cl$	O	1.920(7)	O,O	109.0(4,2.9)	
					H_2O	1.921(9,19)		119.1(5)	
					Cl	2.377(9)	O,Cl	109.2(4,5.6)	

Table 6. Continued 17.

$Na_2[Cu_2(\mu\text{-}NCS)_2(salal)_2](NCS)_2$	tr	6.989(3)	84.88(5)	CuO_2N_2S	salalO	1.924(3,20)	Cu	5.768(1)	O,O	170.8(1)	226
(green)	P-1	10.445(5)	70.13(4)		salalN	1.950(4)			N,N	174.8(2)	
	2	12.692(6)	86.72(6)		µSCN	1.978(4)			O,N	89.9(2,4.1)	
					µNCS	2.933(2)			O,S	93.0(1,4.7)	
									N,S	92.5(1,2.3)	
				Na^IX_n							
$[Cu(\mu\text{-}Br_4obbz)Mn(H_2O)_3]$.	m	9.295(2)		CuO_2N_2Br	O	1.939(6,28)	Mn	5.394(2)	O,O	92.3(3)	75
$2.5H_2O$	$P2_1/c$	17.981(4)	100.00(2)		N	1.928(7,6)			N,N	83.8(3)[d]	
(not given)	4	15.021(2)			Br	2.971(2)			O,N	90.8(3,1.1)[f]	
									O,Br	95.2(2,1.3)	
									N,Br	97.7(2,3.4)	
				$Mn^{II}O_6$	O	2.170(6,9)			O,O	76.9(2)[d]	
						2.204(6)				91.6(2,8.6)	
					H_2O	2.170(8,11)					
$[(H_2O)_4Cu(\mu$-calis[4]arenesulfo-	m	11.955(3)		CuO_6	O	2.44(3,0)			O,O	not given	227
nate)$Na_2].7.5H_2O$	$P2_1$	28.499(6)	91.49(2)		H_2O	1.97(2,0)					
(pale blue)	4	12.442(3)									
				Na^IO_x	O	not given			O,O	not given	
$[Cu(\mu\text{-}ox)_2Na_2(H_2O)_2]$	tr	7.536(3)	81.90(6)	CuO_6	µO	1.930(1,1)	Cu	3.576(2)	O,O	84.5(16,9)[d]	228
(blue)	P-1	9.473(4)	103.77(5)			2.803(2,0)				95.0(7)	
	1	3.576(2)	108.08(4)								
				Na^IO_7	µO	2.464(2)	Na	3.576(2)	O,O	not given	
						2.550(2,37)		3.969(2)			
					H_2O	2.421(2,50)		4.169(2,10)			
$[Cu(\mu\text{-}ox)_2Na_2(H_2O)_2]$	tr	3.583(7)	109.07(1)	CuO_6	µO	1.933(6,1)			O,O	84.5(2,1.6)[d]	229
(deep blue)	P-1	9.649(6)	76.38(1)			2.789(6,0)				95.5(2,1.6)	
	1	7.549(6)	103.32(1)								
				Na^IO_7	µO	2.385-2.587			O,O	88.0-128.3	
					H_2O	2.840(-,5)					

Table 6. Continued 18.

Compound	Crystal class / Space group / Z	a, b, c	α, β, γ	Chromophore	L	M–L	X	M–X	L,L	L–M–L	Ref.
[(H₂O)Cu(μ-ox)₂K₂(H₂O)] (blue)	tr P-1 2	8.7092(7) 10.3904(5) 6.9486(3)	121.110(2) 82.956(2) 110.848(2)	CuO_6	O H_2O	1.947(4,1) 2.462(6,0)			O,O	85.7(2)[d] 90.1(2)	230
				K^IO_8	O H_2O	2.924(7,108) 2.855(7,132)			O,O	not given	
				CuO_6	O	1.943(4,8) 2.631(6,0)			O,O	85.4(2) 89.7(2,8)	
				K^IO_8	O H_2O	2.921(6,279) 3.053(7,0)			O,O	not given	
[Cu(μ-ox)₂Cs₂].2H₂O (blue)	tr P-1 2	9.2980(7) 9.1167(6) 7.1316(4)	97.524(5) 97.410(4) 107.522(4)	CuO_6	O	1.945(11,6) 2.737(14)			O,O	not given	231
				CuO_6	O	1.934(11,0) 2.806(17)			O,O	not given	
				Cs^IO_8	O	3.074(16,15) 3.138(12,38) 3.307(16,86)			O,O	not given	
				Cs^IO_8	O	3.158(16,50) 3.242(13,9) 3.360(13) 3.446(16)			O,O	not given	
[Cu(μ-HCOO)₂Ca(HCOO)₂] (blue)	m P2/c 2	7.300(2) 8.493(2) 6.449(2)	 97.57(1)	CuO_6	μO O	1.982(2,0) 1.964(2,0) 2.525(2,0)	Cu O	3.625(2) 7.295(2) 107.15(1)	O,O	72.79(1) 93.9(1,4.3) 164.0(1,2.7)	232
				$Ca^{II}O_8$	μO O	2.610(2,0) 2.337(2,45) 2.565(2,0)	 O	 138.3(1)	O,O	not given	

Table 6. Continued 19.

Compound				Chromophore							
[Cu(μ-HCOO)$_2$Ca(HCOO)$_4$]	m	22.296(8)		CuO_6	μO	1.965(2,8)	Cu	3.605(2)	O,O	73.54(1)	232
(blue green)	C2/c	8.803(2)	101.00(5)			2.525(2,0)		11.326(2)		93.0(1,3.7)	
	4	6.377(4)					O	106.46(1)		165.6(1,1.1)	
				$Ca^{II}O_7$	μO	2.628(2,0)			O,O	not given	
					O	2.302(2,12)	O	138.0(1)			
						2.498(2,88)					
[Cu(μ-HCOO)$_2$Sr(HCOO)$_2$]	m	7.3296(12)		CuO_6	μO	1.957(3,10)			O,O	not given	233
(blue)	P2/c	8.6402(9)	97.22(1)			2.613(3)					
	4	6.6690(8)			O	1.947(3)					
				$Sr^{III}O_8$	O	not given			O,O	not given	
[Cu(HCOO)$_6$Sr$_2$(H$_2$O)$_8$]	tr	6.61(1)	104.5(1)	CuO_6	O	1.95(1)x2	Sr	3.58(1)	O,O	83.7(12,1.4)	234
(light blue)	P-1	8.84(2)	96.0(1)			2.00(1)x2				93.0(12)	
	1	8.90(2)	88.5(1)			2.45(1)x2					
				$Sr^{II}O_9$	O	2.54(1)	Sr	3.80(1)	O,O	not given	
						2.64(1,3)		4.15(1)			
					H_2O	2.62(1,1)					
						2.76(1,5)					
[(H$_2$O)$_2$Cu(μ$_3$-ox)(μ-ox)$_2$Sr$_2$	tr	6.349(2)	73.21(1)	CuO_6	μ$_3$oxO	1.940(8,28)			O,O	85.9(3,6)[d]	235
(H$_2$O)$_5$]	P-1	10.258(2)	93.66(2)		H_2O	2.499(11,13)					
(blue)	2	15.737(2)	76.44(2)								
				$Sr^{II}O_9$	μO	2.614(5,37)			O,O	72.0(1,5)[d]	
					O	2.553(5,2)					
						2.744(5,92)					
					H_2O	2.688(8,82)					
[Cu$_{0.5}$(μ-HCOO)$_2$Zn$_{0.5}$(H$_2$O)$_2$]	m	8.77(3)		CuO_6	O	2.06(2,40)			O,O	92.5(7,9)	236
(not given)	P2$_1$/c	7.25(2)	98.0(5)			2.22(1)					
	4	9.17(2)									
				$Zn^{II}O_6$	O	2.19(1,0)			O,O	90.0(5,1.3)	
					H_2O	2.07(1,30)					

Table 6. Continued 20.

[(py)$_2$(μ-OH)$_2$Cu$_3$(μ-CrO$_4$)$_2$]	tr	5.378(1)	93.32(1)	CuO$_6$	μHO	1.934(4,0)	Cu	3.052(1,19)	O,O	80.9(1)	237
(light brown)	P-1	5.619(1)	100.25(1)			2.478(4,0)		3.310(1)		86.5(3,4)	
	1	13.569(2)	98.45(1)		μO$_3$CrO	1.982(3,0)					
				CuO$_5$N	μHO	1.941(4)			O,O	79.0(1,4.3)	
						2.418(4,42)				96.2(2,15.1)	
					μO$_3$CrO	1.996(3,27)				165.2(1,5.1)	
									O,N	92.2(3,5.9)	
										161.0(3)	
				CrO$_4$	μO	1.668(4,47)			O,O	109.5(2,1.6)	
					O	1.600(4,0)					
[(MeOH)$_2$Cu(μ-mtm)$_2$Mn(H$_2$O)$_2$]	tr	7.228(1)	102.70(1)	CuO$_6$	O	1.960(1,11)	Mn	5.013	O,O	84.4(1,7)[d]	238
(green)	P-1	7.338(2)	91.17(1)		MeHO	2.482(2)x2		5.317			
	1	11.392(1)	93.37(1)								
				Mn^{II}O$_6$	O	2.196(1)x2			O,O	94.0(1,6)	
						2.215(1)x2					
					H$_2$O	2.185(1)x2					
[(H$_2$O)$_6$Cu$_3$(μ-dgl)$_6$Gd$_2$].3H$_2$O	hx	15.126(3)		CuO$_6$	O	1.956(3)x4	Gd	5.706(7)	O,O	90.0(1,1.0)	239
(green-blue)	P6/mcc				H$_2$O	2.519(4)x2	Cu	7.345		180.0(2)	
	2	14.690(8)									
				Gd^{III}O$_9$	O	2.448(2)x6	Gd	7.345	O,O	61.81(7)	
						2.552(2)x3				79.5(1,5.7)	
										121.8(1,1.8)	
										143.1(1,4.5)	
[(H$_2$O)$_4$Cu(μ-sql)$_2$La(H$_2$O)$_6$	m	6.728(1)		CuO$_6$	O	1.96(1,1)			O,O	90.0(5,7.2)	240
(μ-sql)$_2$Sm(H$_2$O)$_6$].2H$_2$O	P2$_1$	32.117(9)	111.60(1)		H$_2$O	1.99(1,1)				175.4(6,5)	
(yellow)	2	8.150(1)				2.43(1,1)					
				LnO$_9$	O	2.45(2,7)			O,O	61.4-149.8(6)	
					H$_2$O	2.56(2,8)					

Table 6. Continued 21.

{[(μ-sql)La(H_2O)$_6$]$_2$(μ-sql)$_2$Cu	m	6.780(2)		CuO_6	O	1.963(2)x2	La	5.7258(3)	O,O	90.0(1,5.8)	241
(H_2O)$_4$}.2H_2O	$P2_1/c$	32.311(4)	111.55(2)		H_2O	1.977(3)x2	Cu	6.780(2)			
(yellow)	2	8.162(2)				2.426(2)x2					
				$La^{III}O_9$	O	2.462(3,10)	La	6.780(2)	O,O	not given	
						2.536(2)					
					H_2O	2.588(3,49)					
{[(μ_3-sql)Gd(H_2O)$_4$]$_2$(μ-sql)$_2$Cu	m	13.784(1)		CuO_6	O	1.962(2)x2	Gd	5.5642(5)	O,O	90.0(1,1.3)	241
(H_2O)$_4$}.2H_2O	C2/c	8.219(1)	94.20(7)		H_2O	1.964(3)x2		5.6631(5)			
(green)	4	26.268(2)				2.377(3)x2					
				$Gd^{III}O_8$	O	2.411(2,31)	Gd	5.8457(5)	O,O	not given	
					H_2O	2.402(3,44)					
[(en)$_2$Cu(μ-NCS)$_2$Hg(SCN)$_2$]	m	7.51_0		CuN_6	enN	2.07(-,10)			N,N	70.31	242
(deep purple)	$P2_1/c$	18.03_2	96.53(2)		μSCN	2.58-2.81				85.12	
	4	14.20_1									
				$Hg^{II}S_4$	μNCS	2.55(-,6)			S,S	107.3(-,7.2)	
					NCS	2.56(-,3)				120.12	
[(bpy)$_2$Cu(μ-NC)$_2$Ag][Ag(CN)$_2$].	tr	11.302(2)	97.09(2)	CuN_6	bpyN	2.029(3,19)			N,N	79.1(1,1.7)[d]	243
H_2O	P-1	11.814(1)	109.44(2)		μCN	2.003(3)				92.4(1,8.0)	
(not given)	2	10.457(1)	77.89(1)			2.676(5)				170.9(1,1.9)	
				Ag^IC_2	μNC	2.079(5,1)			C,C	180	
				Ag^IC_2	NC	2.077(5,4)			C,C	173.5(2)	
[(en)$_2$Cu(μ-NC)Ni(CN)$_2$].2PhNH$_2$	m	9.547(4)		CuN_6	enN	1.986(5,20)			N,N	87.1(2)[d]	244
(blue)	$P2_1/c$	10.621(5)	107.89(2)		μCN	2.593(5,0)				89.2(2,1.2)	
	2	12.746(3)									
				$Ni^{II}C_4$	μNC	1.906(5,0)			C,C	88.0(3)	
					NC	1.863(6,0)					

Table 6. Continued 22.

$[(H_3N)_2Cu(\mu\text{-}NC)_4Ni].2C_6H_6$	tg	7.345(3)		CuN_6	H_3N	2.05(2,0)	N,N	not given	245
(pale green)	P4/mcc				μCN	2.20(1,0)			
	2	16.519(4)							
				$Ni^{II}C_4$	μNC	1.88(1,0)	C,C	not given	
$[Cu(\mu\text{-}nta)Na(H_2O)]$	or	9.899(5)		CuO_5N	O	1.966(4,23)	O,O	89.8(2,3.8)	246
(pale blue)	$P2_12_12_1$	12.565(5)				2.324(4,27)		118.5(2)	
	4	7.548(5)			N	2.086(5)		163.0(2,10.5)	
							O,N	76.3(2,1.6)[d]	
								88.7(2,3.1)	
								165.7(2)	
				Na^IO_5	O	2.313(5,31)	O,O	81.9(2,2.4)	
						2.370(5,2)		98.0(2,7.1)	
					H_2O	2.288(5)		134.4(2,7)	
								162.6(2)	
$[(H_2O)Cu(\mu\text{-}nta)_2(\mu_3\text{-}nta)K(H_2O)]$	m	6.95(2)		CuO_5N	µntaO	1.98(2)	O,O	90.8(5,4.8)	247
$.H_2O$	$P2_1/b$	17.60(3)				2.34(2)		118.0(5)	
(pale blue)	4	10.00(2)	112.5(5)		N	2.09(3)		152.0(5)	
					$\mu_3$ntaO	1.92(2)	O,N	75(1)[d]	
					H_2O	2.02(2)		88.5(1,4.5)	
								166(1)	
				K^IO_6	µntaO	2.71(3,8)	O,O	not given	
					H_2O	2.78(3)			
$[(H_2O)Cu(\mu\text{-}CN)_5Fe(NO)].H_2O$	or	13.4911(9)		CuN_5O	μCN	1.981(5,5)	N,N	92.7(3,6.2)	248
(green)	Pnma	7.1698(6)				2.160(7)		163.7(2)	
	4	10.8623(7)			H_2O	2.580(7)	N,O	81.8(3,4.0)	
								175.5(4)	
				$Fe^{III}C_5N$	μNC	1.929(2,6)	C,C	87.9(3,3.2)	
					ON	1.658(7)		170.2(2)	
							C,N	94.9(2,7)	
								179.7(5)	

Table 6. Continued 23.

[Cu(μ-edta)Mn(H_2O)$_4$].2H_2O	or	14.626(3)		CuO_4N_2	O	1.991(5,25)			O,O	108.6(2)	249
(not given)	Pn2$_1$a	13.068(4)				2.252(4,37)					
	4	9.969(3)			N	2.043(5,1)					
				$Mn^{II}O_6$	O	2.149(7,13)			O,O	not given	
					H_2O	2.195(9,51)					
[(H_2O)Cu(μ-edta)Ag].H_2O	m	8.3490(7)		CuO_4N_2	O	1.938(4,5)			O,O	58.4(2)	250
(not given)	P2/n	17.725(2)	90.866(9)			2.524(4)				92.6(2,7)	
	4	10.129(9)			N	2.073(4)				166.1(2,6.5)	
						2.344(4)			N,N	93.5(2)	
					H_2O	1.994(4)			O,N	79.8(2)[d]	
										93.3(2,2.8)	
										115.7(2,1.3)	
										153.6(1)	
				Ag^IO_4	O	2.247(4,10)			O,O	91.8(2,5.2)	
						2.432(5,67)				113.0(2,1)	
										154.0(2)	
[Cu(μ-trisH)(μ-tris)Na](ClO_4)$_2$	tr	9.946(1)	109.39(1)	CuO_4N_2	O	1.967(2,1)	Na	3.138(2)	O,O	77.23(6)	251
(blue)	P-1	11.793(2)	112.72(1)			2.475(3,62)		4.067(2)		87.67(5)	
	2	8.853(2)	82.02(1)		N	1.999(3,2)				106.67(9)	
										172.9(1)	
									N,N	101.9(1)	
									O,N	84.43(9)[d]	
										174.1(1)	
				Na^IO_6	O	not given			O,O	not given	

Table 6. Continued 24.

Compound				Chromophore					Ref.
[Cu(μ-pdac)$Na_2(H_2O)_4$]	m	24.493(5)		CuO_4N_2	O	1.927(2,1)	O,O	87.78(6,4.30)	252
(blue)	$P2_1/n$	8.222(1)	90.43(1)			2.136(2,15)		136.08(6)	
	4	9.805(1)			N	2.260(2,7)		167.77(6)	
							N,N	77.20(6)[d]	
							N,O	76.07(6,28)[d]	
								82.25(6,37)	
								107.28(6,1.30)	
								145.57(6,28)	
				Na^IO_x	O	not given	O,O	not given	
[Cu(μ-edds)$Mg(H_2O)_4$].$3H_2O$	or	8.392(2)		CuO_4N_2	O	1.984(4,3)	O,O	92.5(2,10.2)	253
(blue)	$P2_12_12_1$	10.855(2)				2.304(4)		169.7(1)	
	4	21.508(6)			μO	2.546(4)	N,N	85.6(2)	
					N	2.007(5,6)	O,N	74.7(2,3.1)[d]	
								95.7(2,5.6)	
								173.3(2,3.0)	
				$Mg^{II}O_6$	O	2.075(5)	O,O	90.0(2,4.7)	
					μO	2.109(5)		176.6(2,2.7)	
					H_2O	2.077(5,8)			
[Cu($μ_3$-edta)$Ca(H_2O)_2$].$2H_2O$	m	8.596(5)		CuO_4N_2	O	1.982(6,26)	O,O	90.6(1,3.7)	254
(not given)	$P2_1/b$	17.486(8)				2.711(6)	N,N	84.0(1)[d]	
	4	11.024(8)	97.14(2)		N	2.076(5)	O,N	73.5(1,4.9)[d]	
						2.344(5)		89.4(1,7.5)	
								103.0(1,1.6)	
				$Ca^{II}O_6$	O	2.296(3)	O,O	86.8(2,3.6)	
						2.318(3,0)		93.2(2,3.5)	
					H_2O	2.360(3,22)		175.0(2)	

Table 6. Continued 25.

[Cu(μ-edta)Ca(H_2O)$_2$].2H_2O (blue)		not given		CuO_4N_2	μedtaO μedtaN	1.982(-,26) 2.709(-) 2.076(-) 2.344(-)	O,O	106.9(-)	255
				$Ca^{II}O_6$	μedtaO H_2O	2.355(-,40) 2.433(-,20) 2.133(-) 2.395(-)	O,O	not given	
[Cu(μ-edta)Sr(H_2O)$_3$].H_2O (blue)	m $P2_1/c$ 4	8.717(1) 10.568(3) 19.374(3)	100.39(1)	CuO_4N_2	μedtaO N	1.970(4,17) 2.305(4) 2.084(5) 2.449(4)	O,O	92.6(2)	255
				$Sr^{II}O_8$	μedtaO H_2O	2.577(6,44) 2.638(7,48)	O,O	not given	
[Cu(μ-edta)Ba(H_2O)$_3$].H_2O (blue)	m $P2_1/c$ 4	8.818(1) 10.200(2) 19.737(2)	102.40(1)	CuO_4N_2	μedtaO N	1.961(3,35) 2.355(4) 2.056(3) 2.495(3)	O,O	98.0(1)	255
				$Ba^{II}O_8$	μedtaO H_2O	2.756(4,68) 2.832(4,99)	O,O	not given	
[(H_2O)$_3$(py)$_2$Cu(μ-calix[4]arene sulfonate)Na_3].13H_2O (not given)	m $P2_1/n$ 4	14.672(3) 25.577(5) 15.457(3)	101.94(4)	CuO_4N_2	O H_2O pyN	2.55(1) 2.01(1,1) 2.50(1) 1.99(1,1)		not given	256
				$Na^{I}O_5$	O	not given		not given	
				$Na^{I}O_6$	O	not given		not given	

Table 6. Continued 26.

[(dmso)$_2$Cu$_3$(μ-opba)$_3$Mn$_2$(μ-rad)] .2H$_2$O (green)	m Cc 4	25.379(3) 25.146(3) 18.845(6)	131.52(4)	CuO$_4$N$_2$ (x2)	opbaO 1.96(2,3) opbaN 1.95(3,1) radO 2.92(2,2)	Cu 8.214 10.356	not given	257
				CuO$_4$N$_2$	opbaO 2.03(2,3) N 1.95(2,3) dmsoO 2.56(6,1)		not given	
				Mn^{II}O$_6$	opbaO 2.23(2,3) radO 2.18(2,3)		not given	
[(pyor)$_4$Cu(μ-Cl)$_2$Pd$_2$(μ-Cl)$_2$Cl$_2$] (dark brown)	m P2$_1$/c 2	10.8184(19) 13.3823(16) 10.4782(20)	115.780(14)	CuO$_4$Cl$_2$	O 1.965(4,22) Cl 2.823(1,0)	Cl 140.06(5)	O,O 93.44(5) O,Cl 89.6(1,1.2)	258
				Pd^{II}Cl$_4$	Cl 2.264(1) μCl 2.271(1) 2.330(1,1)		Cl,Cl 90.0(1,3.5)	
[(py)$_4$Cu(μ-F)$_2$W(O)$_2$F$_2$] (blue) (at 153.2 K)	m C2/c 4	10.450(2) 13.396(3) 16.067(7)	97.33(2)	CuN$_4$F$_2$	pyN 2.042(6,29) μF 2.364(5)	F 174.9(3)	N,N 87.1(2) N,F 89.8(2)	259
				W^{VI}F$_4$O$_2$	μF 1.891(4,0) X 1.869(5,14)		X,X 89.8(2) 180.0	
[(py)$_2$(H$_2$O)$_2$Cu(μ-O)(μ-F)WOF$_3$] (blue) (at 153.2 K)	m P2$_1$/n 4	11.500(4) 7.970(2) 16.521(3)	103.06(2)	CuO$_3$N$_2$F	μX 2.360(7,4) H$_2$O 1.997(7,13) pyN 2.022(9,12)	X 140.4(5,3)	O,O 177.1(3) O,X 89.3(4) X,X 179.2(2) O,N 88.7(3,1.5) N,X 90.5(5) N,N 179.0(2)	259
				W^{VI}X$_6$	μX 1.865(6,11) X 1.872(6,2) 1.926(6)		X,X 90.4(3,2.5) 177.8(3,6)	

Table 6. Continued 27.

[(H$_2$O)Cu(μ$_3$-or)(μ-gly)Na(H$_2$O)$_3$]	m	7.7477(15)		CuN_4O_2	glyN	1.981	N,N	101.4	260
(blue)	P2$_1$	7.2138(17)	105.94(2)		μglyO	1.951	O,O	87.9(-,2.1)	
	2	13.1318(34)			orN	1.981	N,O	83.6(-,1.1)[d]	
					μ$_3$orO	2.731		94.4(-,7)	
					H$_2$O	2.480		169.9(-,2)	
				$Na^{I}O_6$	glyO	2.397(-,26)	O,O	not given	
					μorO	2.421(-,14)			
					μ$_3$orO	2.586			
					H$_2$O	2.356			
[(MeCN)Cu(μ-[9]aneN$_3$)Na	m	9.6373(22)		CuN_4O_2	N	2.008(3,7)	N,N	86.95(12)[d]	261
(BF$_4$)$_2$].NO$_3$	P2$_1$/a	30.814(12)	114.433(22)			2.611(5)		97.85(12)	
(blue)	4	10.1042(22)			O	1.963(3)		168.40(12)	
						2.2518(25)	N,O	79.9(1,2.2)[d]	
					MeCN	1.984(3)		96.57(10)	
								169.07(11)	
				$Na^{I}F_5O$	F	2.242(3)-		not given	
						2.424(3)			
					O	2.321(3)			
[Cu(μ-pvoH)(μ-pvo)K]	tr	11.547(5)	71.17(3)	CuN_4O_2	N	1.95(1,1)	N,N	82.6(5,7)[d]	262
(dark red)	P-1	10.625(4)	116.87(3)		O	2.58(1)		96.7(5,2.4)	
	4	10.328(4)	114.02(3)			2.97(1)	O,O	174.4(4)	
							N,O	90.0(5,9.2)	
				CuN_4O_2	N	1.93(1,2)	N,N	83.2(5,1.9)[d]	
					O	2.71(1)		97.0(6,8)	
						2.83(1)	O,O	163.9(4)	
							N,O	89.9(5,11.1)	
				$K^{I}O_3N_3$	O	2.71(1,9)		not given	
					N	2.81(1,11)			
				$K^{I}O_4N_3$	O	2.81(1,28)		not given	
					N	2.88(1,13)			

Table 6. Continued 28.

Compound	Cryst. cl., space group, Z	a, b, c	β	Chromophore	M–L	M…M	L–M–L	Ref.
$[(en)_2Cu(\mu\text{-}Cr_2O_7)]$ (yellow brown)	m C2/c 4	16.166(2) 5.6830(9) 15.106(2)	109.45(1)	CuN_4O_2	enN 2.012(4,4) μO 2.525(3,0)		N,N 85.0(1)[d] 95.0(1) 180 O,O 180 N,O 89.3(1,3.4)	263
				$Cr^{VI}O_4$	O 1.602(3,6) 1.790(2) μO 1.627(3)		O,O 109.5(2,2.2)	
$[(H_2O)_2((en)_2Cu)_2((\mu\text{-}edta)Fe)_2$ $(\mu\text{-}O)]$ (red purple)	m C2/c 8	16.662(2) 15.254(3) 18.751(3)	102.58(1)	CuN_4O_2	enN 2.011(3,8) H_2O 2.652(3)	Fe 5.734(1) 5.986(1) 6.681(1)	N,N 84.7(1)[d] N,O 86.1(1,3.2)	264
				CuN_4O_2	enN 2.002(3,3) O 2.678(3)		N,N 83.6(1)[d] N,O 94.5(1,8.0)	
				$Fe^{III}O_4N_2$	μO 1.7735(5) O 2.024(3,15) N 2.236(3,10)	Fe 3.533(1) O 169.7(2)	O,O 96.8(1,7.8) 162.63(9) N,N 80.2(1)[d] O,N 78.1(1,1.6)[d] 92.1(1,8.8) 154.3(1)	
$[(tact)Cu(\mu\text{-}Cl)CdCl(H_2O)_2]Cl$ (red violet)	m C2/c 4	15.819(8) 13.233(7) 9.240(5)	94.31(5)	CuN_4Cl_2	tactN 2.006(3,5) μCl 2.918		N,N 85.62(14)[d] 180.0	30
				$Cd^{II}Cl_4O_2$	Cl 2.691(1) μCl 2.582(2) H_2O 2.326(3)		Cl,Cl 89.15(5) 180.0 O,Cl 89.2(1,5) 180.0	
$[(en)_2Cu(\mu\text{-}Cl)_2Pt(en)_2](ClO_4)_4$ (red)	or Icma 4	13.645(1) 10.787(1) 9.645(1)		CuN_4Cl_2	enN 2.031(4,0) μCl 3.081(3,0)		N,Cl 91.2(1)	265
				$Pt^{IV}N_4Cl_2$	enN 2.313(3,0) μCl 3.081(3,0)		not given	

Table 6. Continued 29.

$[(en)_2Cu(\mu\text{-}Br)_2Pt(en)_2](ClO_4)_4$	or	13.599(1)		CuN_4Br_2	enN	2.043(3,0)			N,Br	90.9(1)	265
(red)	Icma	10.917(1)			μBr	3.005(1,0)					
	4	9.657(1)									
				$Pt^{IV}N_4Br_2$	enN	2.453(1,0)				not given	
					μBr	3.005(1)					
$[(en)_2Cu(\mu\text{-}Br)_2Pt(en)_2](ClO_4)_4$	or	13.625(2)		CuN_4Br_2	enN	2.043(6,0)			N,Br	90.9(1)	
(red)	Icma	10.905(2)			μBr	3.016(2,0)					
	4	9.651(1)									
				$Pt^{IV}N_4Br_2$	enN	2.436(2,0)				not given	
					μBr	3.016(2,0)					
$[Cu(\mu\text{-}NCS)_4Hg]$				CuN_4S_2	μSCN	1.898(47,48)			N,N	74	266
(not given)	Pcan				μNCS	3.000(17,0)				106	
									M,S	97.9(-,9)	
				$Hg^{II}S_4$	μNCS	2.520(14)			S,S	100.9(3,7.6)	
						2.620(14)				143	
$[(hmpa)_2Cu(\mu\text{-}Cl)_2(Pd(\mu\text{-}Cl)_2)_2]$	m	17.675(3)		$CuCl_4O_2$	μCl	2.350(4,1)			Cl,Cl	77.8(2,1)	267
(not given)	$P2_1/n$	9.932(6)	107.34(2)			2.783(5,12)	Cl	94.7(2,6.0)	O,O	90.9(5)	
	4	18.457(6)			hmpaO	1.91(1,1)					
				$Pd^{II}Cl_4$	μCl	2.288(5,34)			Cl,Cl	90.0(2,3.0)	
$(H_2O)Cu(\mu\text{-}ghg)NaClO_4$	or	12.92		CuO_3N_3	O	2.03				not given	268
(violet)	$P2_12_12_1$	19.34				2.76					
	4	6.62			N	1.96(-,5)					
					H_2O	2.74					
				Na^IO_x	O	not given				not given	

Table 6. Continued 30.

[((hfac)$_2$Cu)$_2$(μ-cphi)$_2$(μ-I)$_2$Pd]	m	8.317(2)		CuO_4NI	O 1.921(4)		O,O 87.6(2,3)	269
(green)	P2$_1$/c	13.541(1)	100.45(1)	(x2)	1.936(4,7)	I 99.81(2)	92.4(2,1)[f]	
	2	22.568(5)			N 2.669(6)		178.5(2,1.2)	
					μI 3.1929(9)			
				$Pd^{II}C_2I_2$	C 1.944(6,0)		I,I 180.00	
					μI 2.5795(6,0)		C,C 180.00	
							I,C 88.4(2)	
[(ddc)Cu(μ-Cl)$_2$ZnCl$_2$]	m	8.812(2)		$CuO_2N_2Cl_2$	ddcO 2.066(5,0)		O,O 79.9(3)[d]	270
(dark green)	P2$_1$/m	15.972(3)	114.33(1)		ddcN 1.987(6,0)	Cl 106.9(1)	N,N 97.9(3)[f]	
	2	8.831(1)			μCl 2.741(3,31)	175.6(2)	Cl,Cl 169.7(1)	
							O,N 91.2(3)[f]	
							O,Cl 86.1(2,5)	
							N,Cl 93.4(2,2)	
				$Zn^{II}Cl_4$	Cl 2.274(2,2)		Cl,Cl 109.5(1,3.2)	
					μCl 2.238(3,0)			
[Br$_2$Cu(μ-gly)$_2$(μ-glyH)$_2$Li$_2$].2H$_2$O	m	9.011(1)		$CuO_2N_2Br_2$	O 1.984(3,0)		O,N 97.1(2)	271
(royal blue)	P2$_1$/n	5.061(1)	91.54(5)		N 1.956(4,0)		O,Br 91.4(1)	
	2	20.659(3)			Br 2.987(1,0)		N,Br 90.5(1)	
				Li^IO_2	O 1.949(6,48)		O,O not given	
[Cu(μ-bet)$_4$Li](ClO$_4$)$_3$	tg	12.096(3)		CuO_8	O 1.970(5)x4	Li 3.116(1)	O,O 54.2(6)[c]	272
(deep blue)	P-42m				μO 2.706(5)x2		90.1(2)	
	1	6.2330(7)					176.9(5)	
				Li^IO_4	μO 1.902(6,0)		O,O 104.5(2)	
							119.1(3)	
[Cu(μ-ac)$_4$Ca(H$_2$O)$_2$].4H$_2$O	tg	11.183		CuO_8	O 1.973(2,0)		O,O 65.7(1)	273
(not given)	I4/m				μO 2.790(3,0)		90.6(1)	
	4	16.277		$Ca^{II}O_6$	μacO 2.295(3,0)		O,O 90.0(1,7.5)	
					H$_2$O 2.358(3,0)			

Table 6. Continued 31.

Compound	Crystal data	Cell	Chromophore	M–L	M–M	L–M–L	Ref.
$[Cu(\mu\text{-}ac)_4Ca(H_2O)_2].4H_2O$	tg	11.152(2)	CuO_8	O 1.969(1,0)	Ca 4.060(1)	O,O 90.6(1)	274
(not given)	I4/m			μO 2.788(2,0)		168.1(1)	
	4	16.240(1)					
			$Ca^{II}O_6$	μacO 2.289(2,0)		O,O 90.0(1,7.3)	
				H_2O 2.343(2,0)			

a Where more than one chemically equivalent distance or angle is present the mean value is tabulated. The first number in parentheses is the e.s.d.,and the second is the maximum deviation from mean value.
b The chemical identity of coordinated atom or ligand is given in the columns.
c Four - membered metallocyclic ring.
d Five - membered metallocyclic ring.
e There are two crystallographically independent molecules.
f Six - membered metallocyclic ring.

Table 6 A. Summary of Cu - M and M - M distances [Å] in heteropolynuclear copper(II) compounds[a].

Cu - M	Distances [ref] M = non transition
- Na(I)	3.138(2), 4.067(2) [251]; 3.576(2) - 4.169(2) [228] 5.698(8,12) [207];
- Sr(II)	3.58(1) [234]
- Hg(II)	3.669(4,43) [193]
	M = transition
- Fe(III)	3.033(2) [200]; 5.04 [213]; 5.734(1), 5.986(1), 6.681(1) [264]
- Mo(VI)	3.452(5,17), 3.533(5,3), 4.289(5), 4.722(5) [212]
- Mn(II)	4.816, 5.506 [199]; 5.013, 5.317 [238]; 5.394(2) [75]; 5.396(1,6) [217]; 5.398(1,4) [218]; 5.412, 6.977 [215]; 5.452(2), 6.066(2) [219]; 5.751 [216]
	M = lanthanoide
- Ce(III)	3.288(2), 3.325(2) [220]
- Gd(III)	5.2799(5) - 7.236(11) [206]; 5.5642(5) [241]
- Tm(II)	5.657(1,7) [203]
- Dy(III)	5.658(2,57) [222]
- Sm(III)	5.672 - 5.742 [221]
- La(III)	5.7258(3) [241]
Cu(II) - Cu(II)	2.633(1) [186]; 2.951(3) [212]; 3.052(1,9), 3.310(1) [237] 3.159(7) - 4.832(7) [193]; 3.576(2) [228]; 3.615(2,10), 7.295(2), 11.326(2) [232]; 3.681(1) [192]; 3.6847(25) [206]; 4.52 [194]; 5.033(1) [217]; 5.118(2) [219]; 5.279 [199]; 5.28 [213]; 5.6631(5), 6.780(2) [241]; 5.74, 6.92 [184]; 5.768(1) [226]; 5.94, 6.51 [179]; 6.367(4), 7.206(4) [207]; 6.545 [216]; 7.505(1) [218]; 8.214, 10.356 [257]
Li(I) - Li(I)	2.529(8) [181]
Mo(VI) - Mo(VI)	3.528(4) [212]
Pt(IV) - Pt(IV)	3.554(2), 3.855(2) [191]
Sr(II) - Sr(II)	3.80(1), 4.15(1) [234]
Na(I) - Na(I)	3.860(5,83) [195]
Mn(II) - Mn(II)	4.819(2) [219]; 5.030(1) [201]; 5.033(1) [217]; 6.997 [216]; 7.209(2) [218]
Gd(III) - Gd(III)	5.8457(5) [241]; 6.336(1) [207]
La(III) - La(III)	6.780(2) [241]
Sm(III) - Sm(III)	11.357 [221]
Dy(III) - Dy(III)	11.939(1) [222]

[a] The first number in parenthesis is e.s.d., the second is maximum deviation from the mean value.

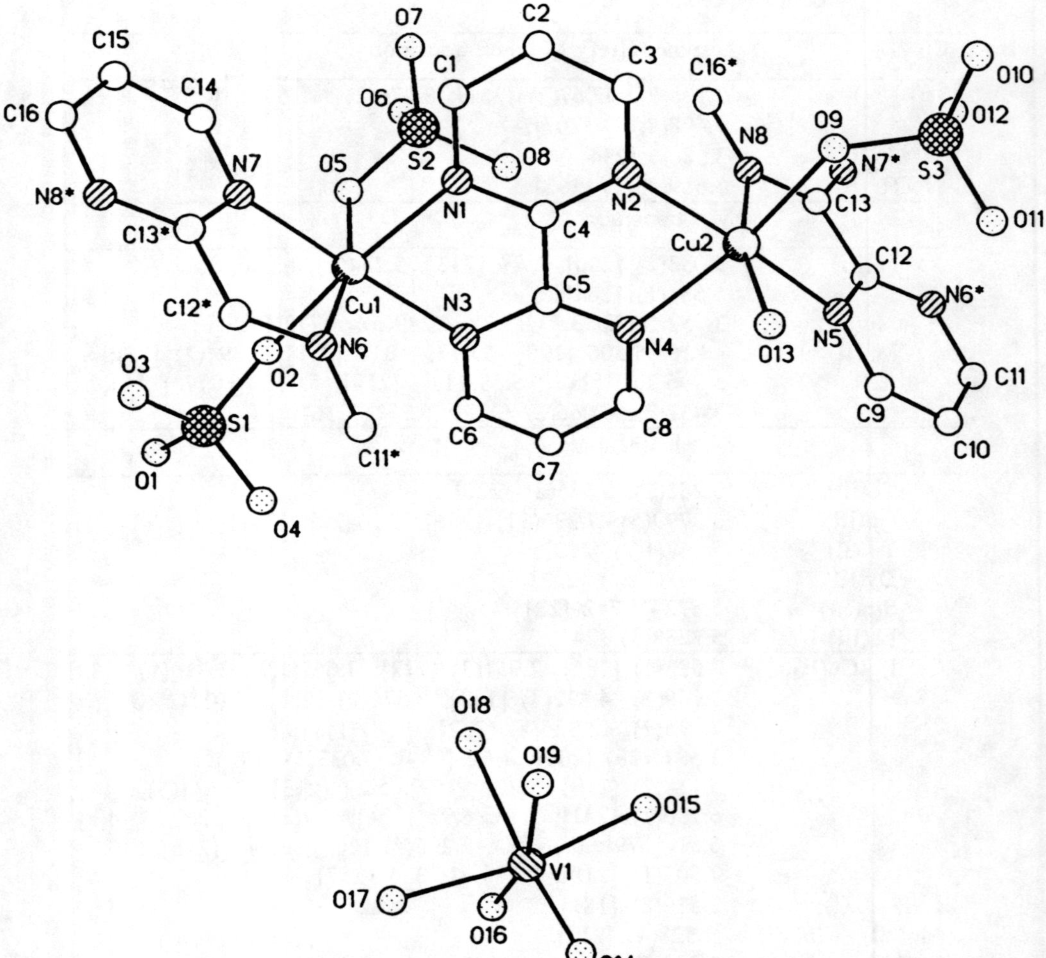

1. Structure of the $[Cu_2(\mu\text{-bipym})_2(H_2O)(SO_4)_3]^{2-}$ anionic chain [29a].

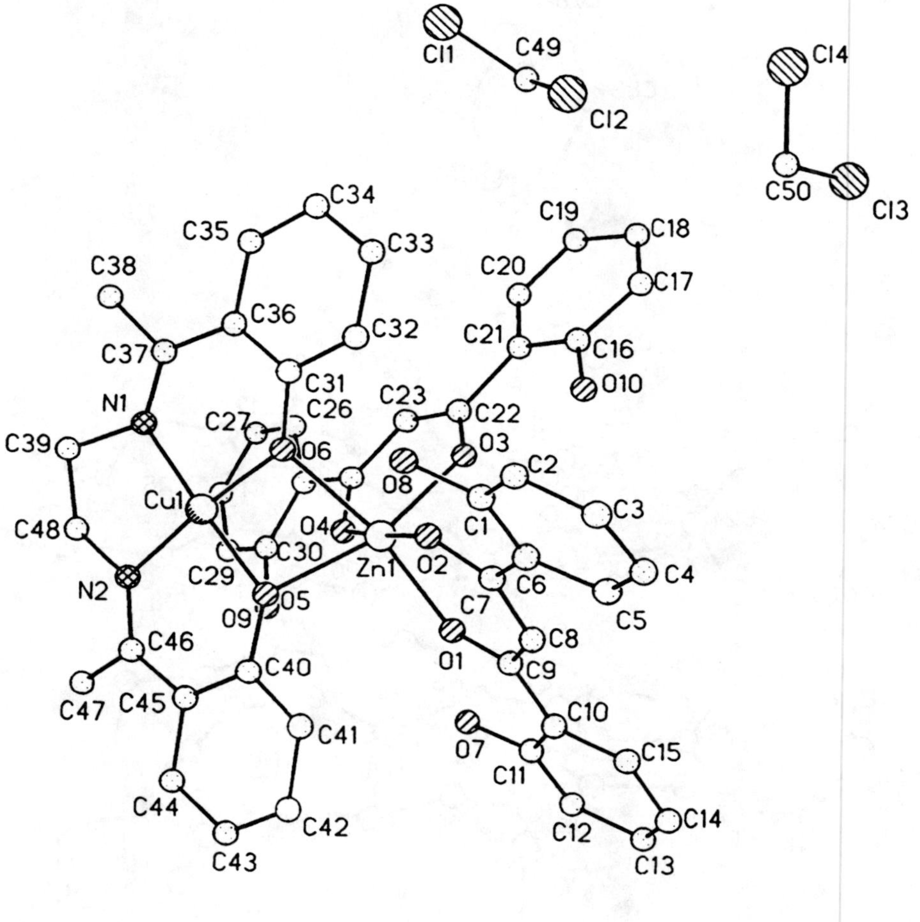

2. Structure of the [Cu(μ-mesalen)Zn(bbppH$_2$)$_2$].2CH$_2$Cl$_2$ [37].

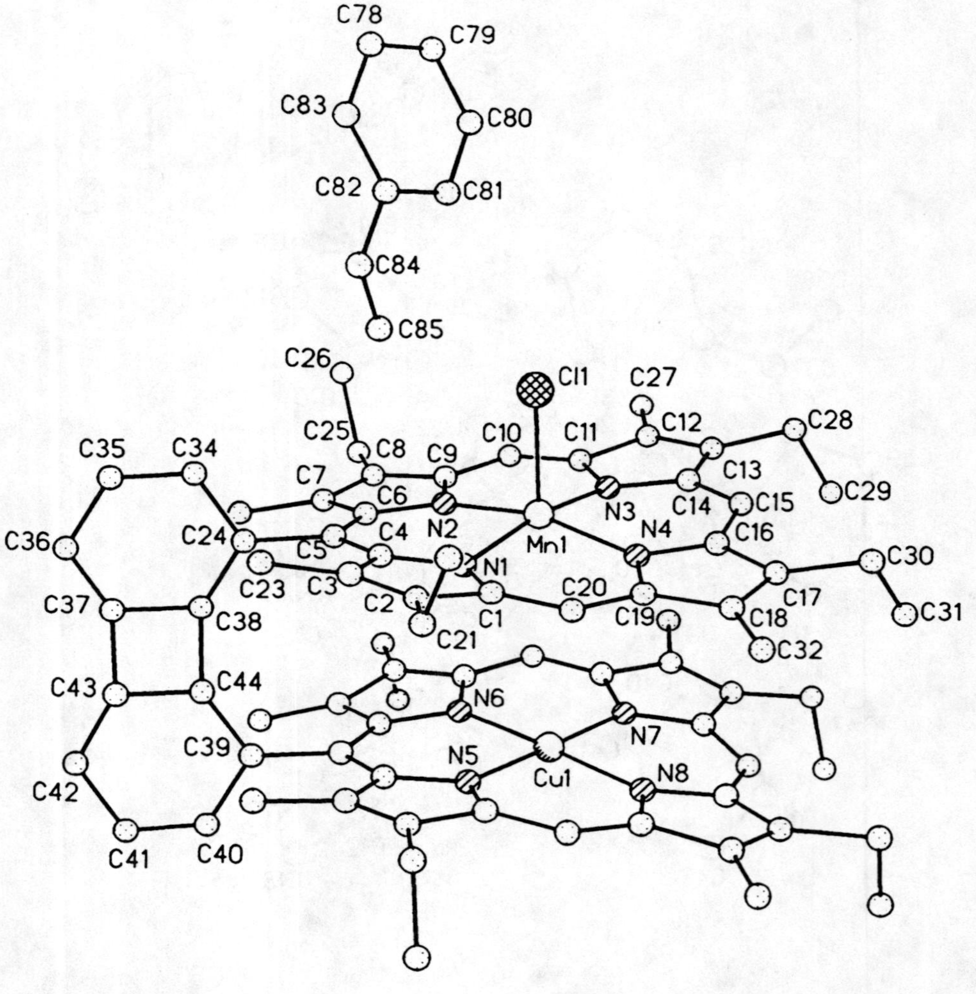

3. Structure of the [Cu(μ-dpb)(MeCl)]C_8H_{10} [72b].

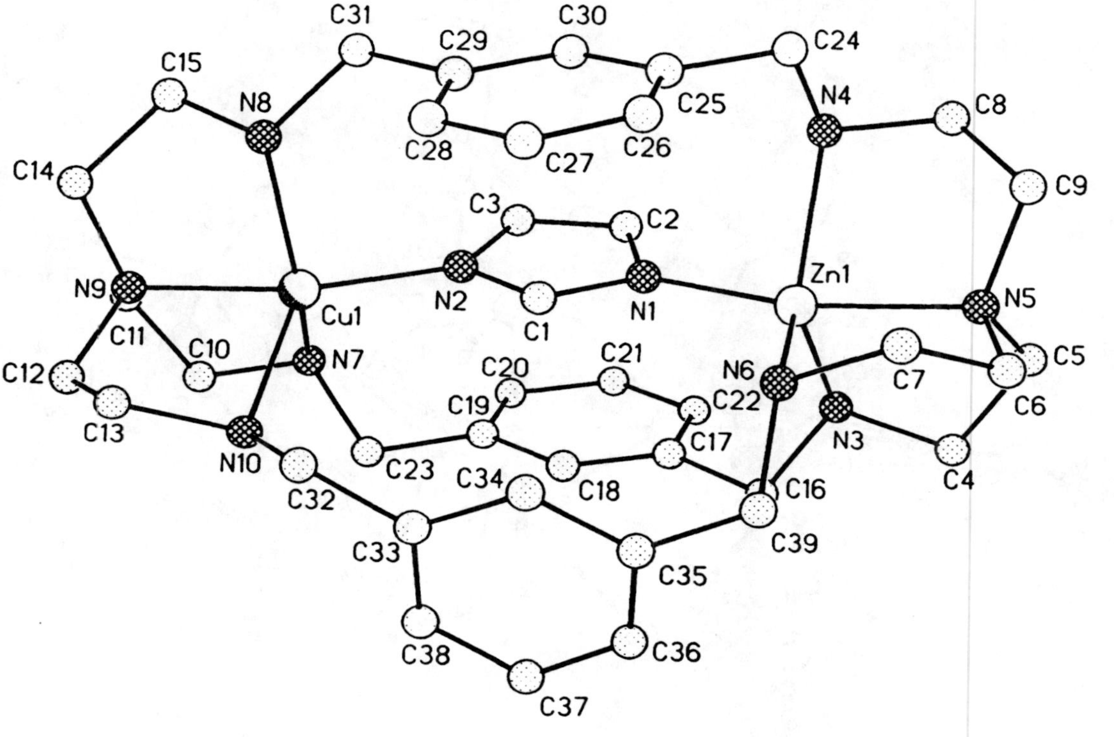

4. Structure of the $[Cu(\mu\text{-im})(\mu\text{-cryptand})Zn]^{3+}$ (a); schematic representation (b) [85].

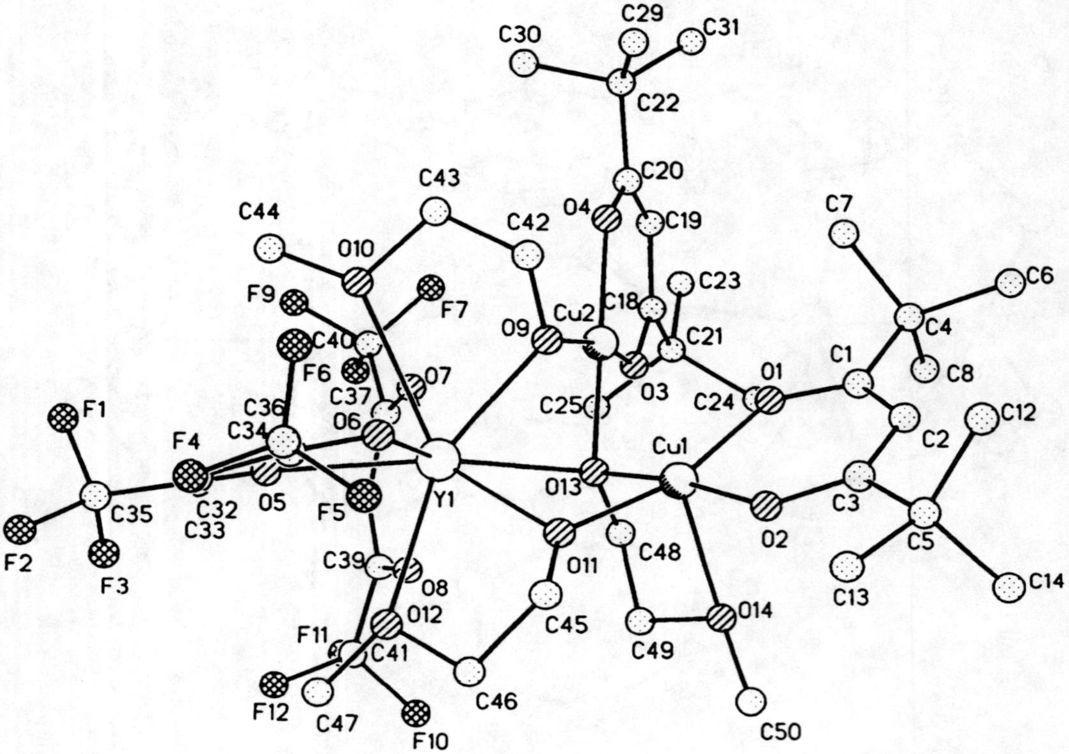

5. Structure of $thd_2Cu_2(\mu\text{-mol})_3Y(hfac)_2$ [107].

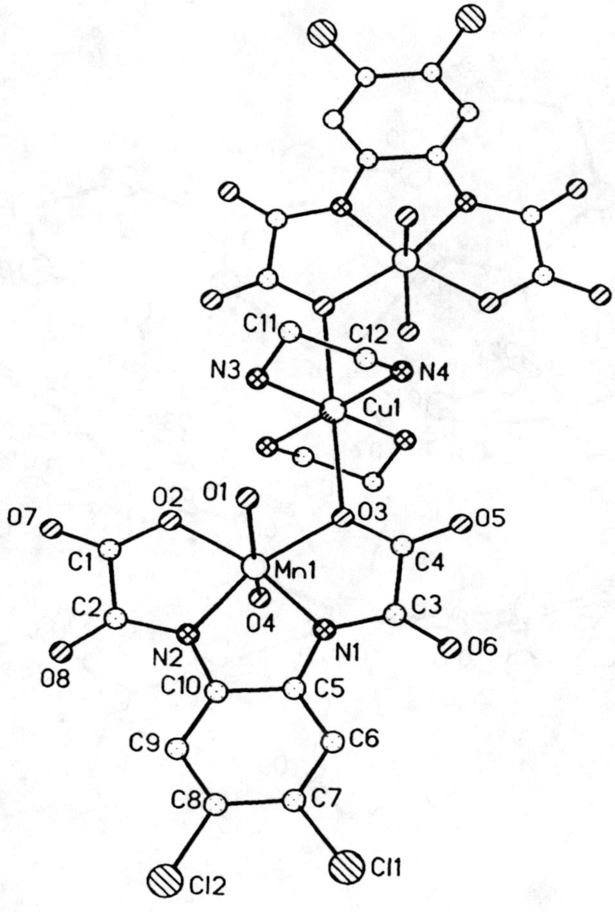

6. Structure of $(en)_2Cu\{(\mu\text{-}Cl_2opba)Mn(H_2O)_2\}_2$ [133].

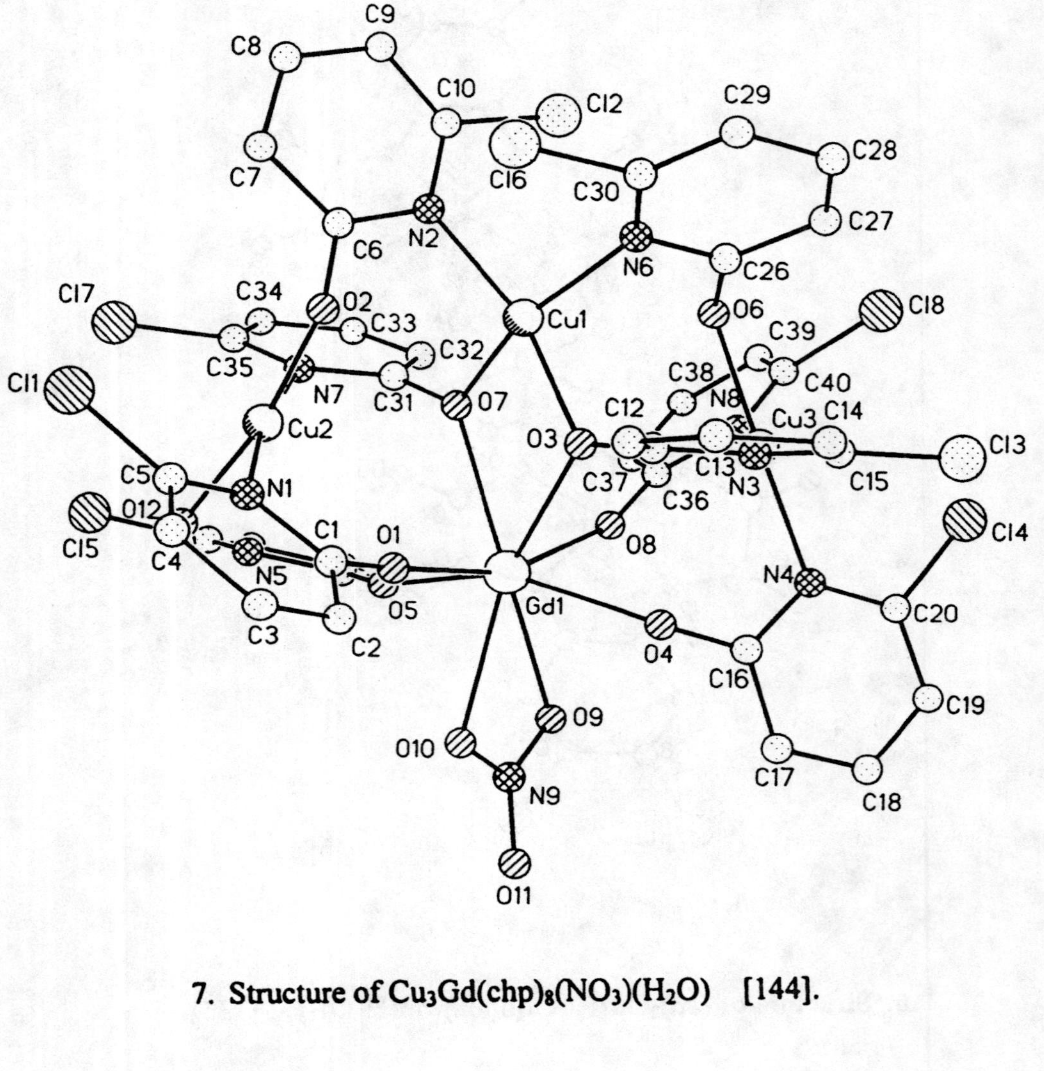

7. Structure of $Cu_3Gd(chp)_8(NO_3)(H_2O)$ [144].

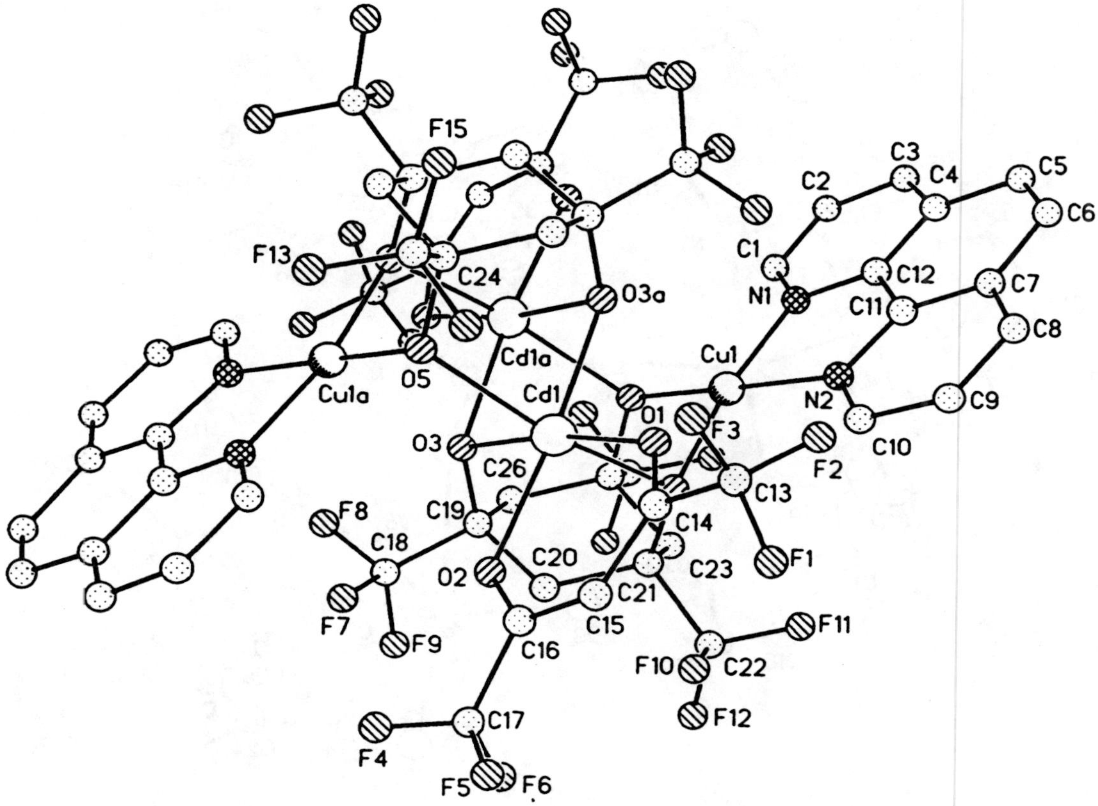

8. Structure of [(phen)$_2$Cu$_2$(μ-ttftch)$_2$Cd$_2$(hfac)$_2$] [154].

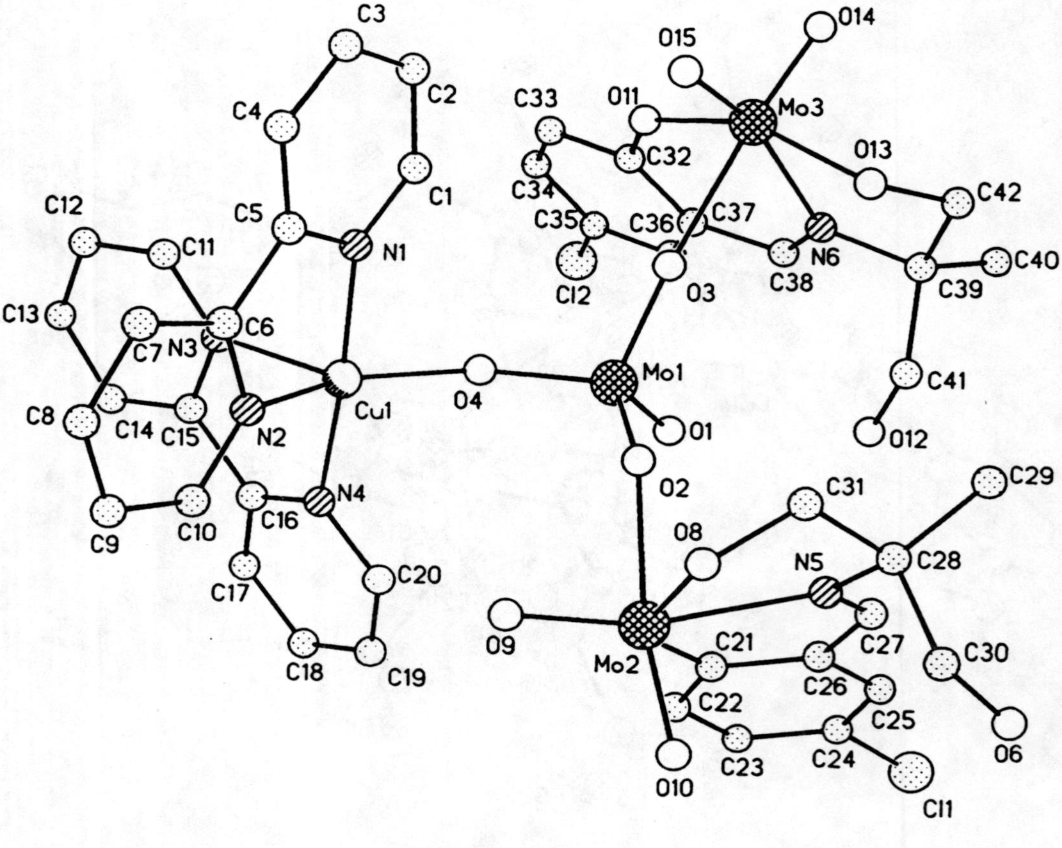

9. **Structure of $CuMo_3O_8(csh)_2(bpy)_2$** [158].

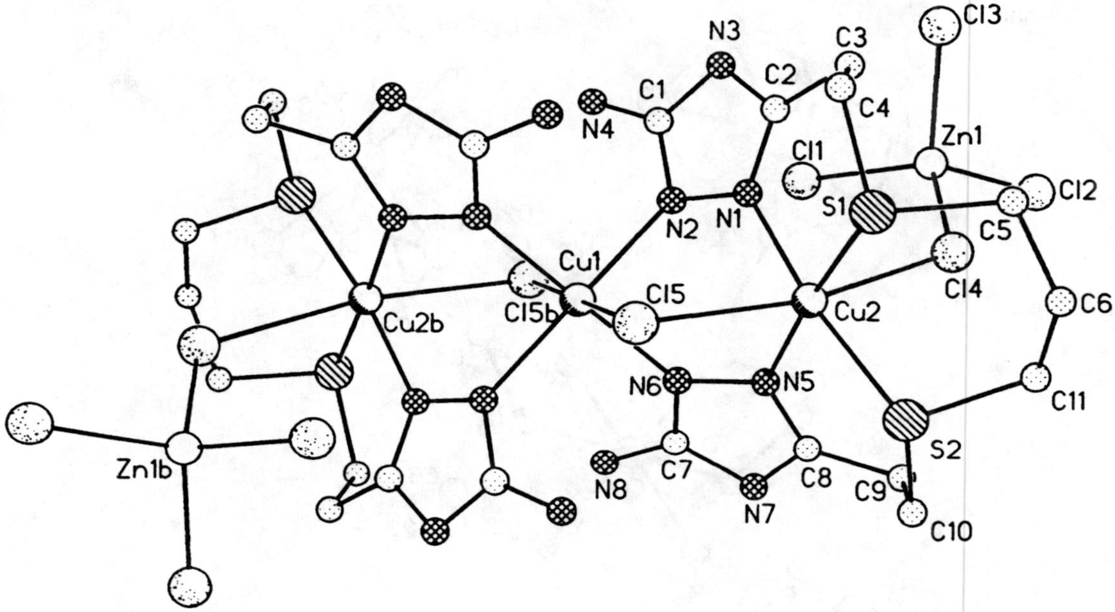

10. Structure of $Cu_3(attn)_2(ZnCl_4)_2Cl_2$ [162].

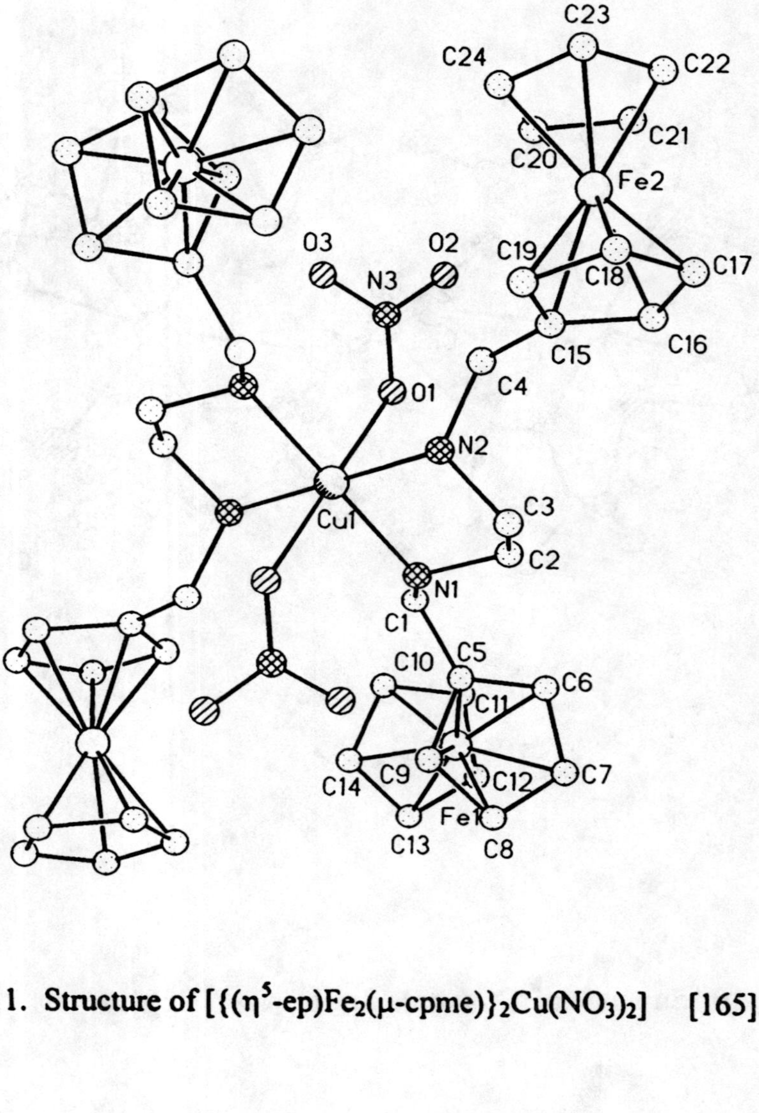

11. Structure of $[\{(\eta^5\text{-ep})Fe_2(\mu\text{-cpme})\}_2Cu(NO_3)_2]$ [165].

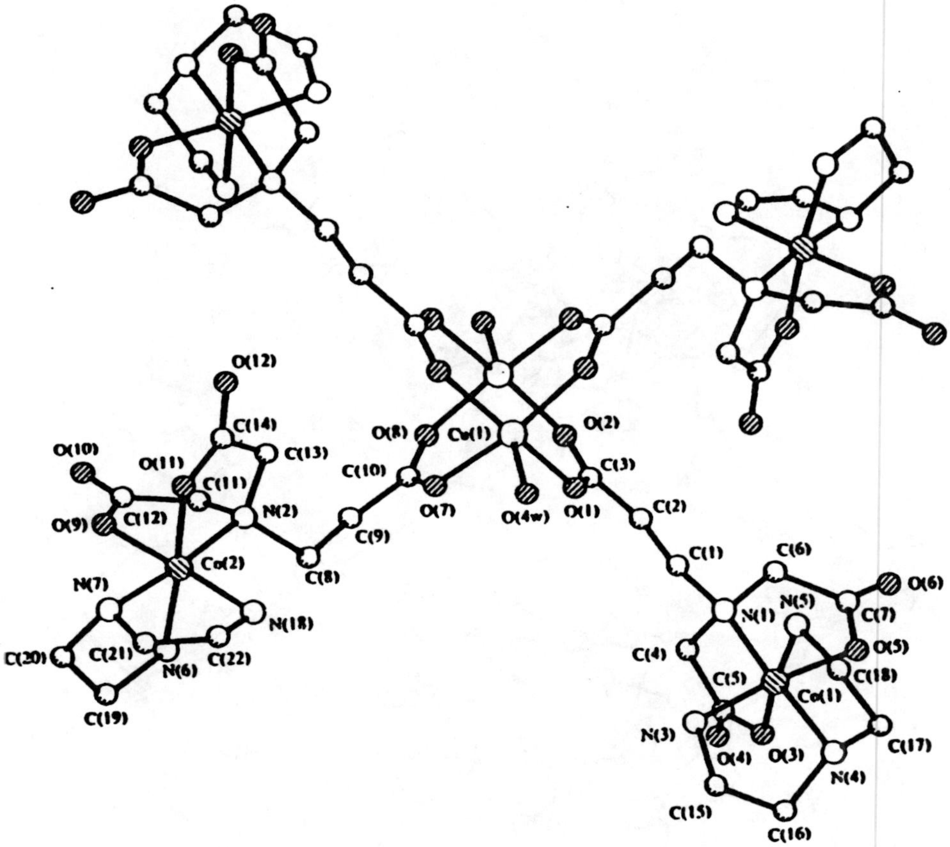

12. Structure of $[\{Co(dien)(cnda)\}_2Cu(H_2O)_2]_2^{4+}$ [170].

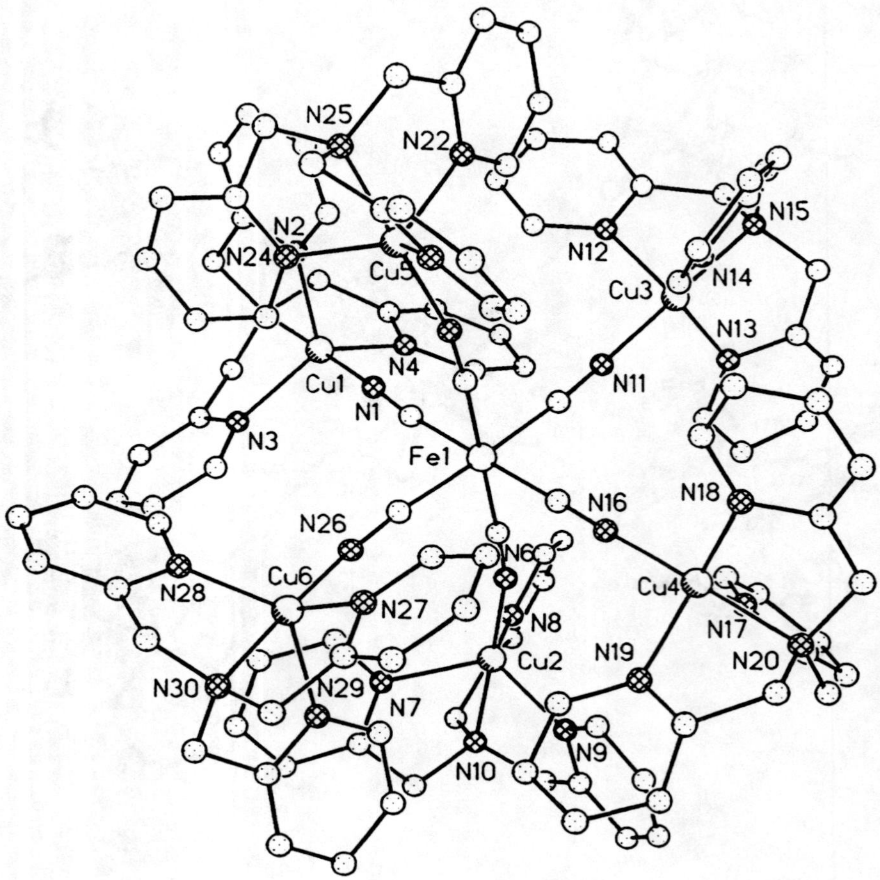

13. Structure of $Fe\{(\mu\text{-}CN)Cu(tpa)\}_6]^{8+}$ [172].

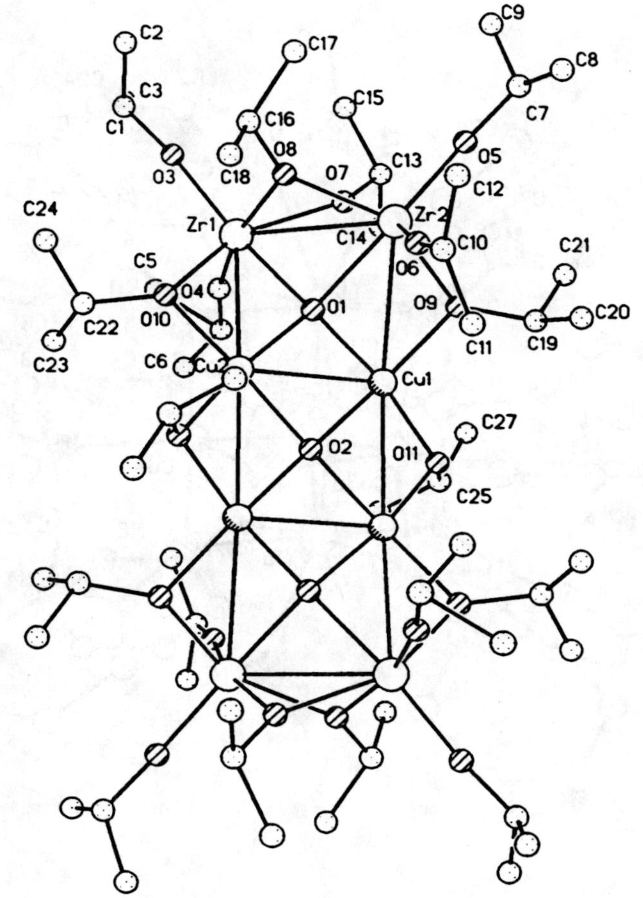

14. Structure of $Cu_4Zr_4O_3(pr^iO)_{18}$ [173].

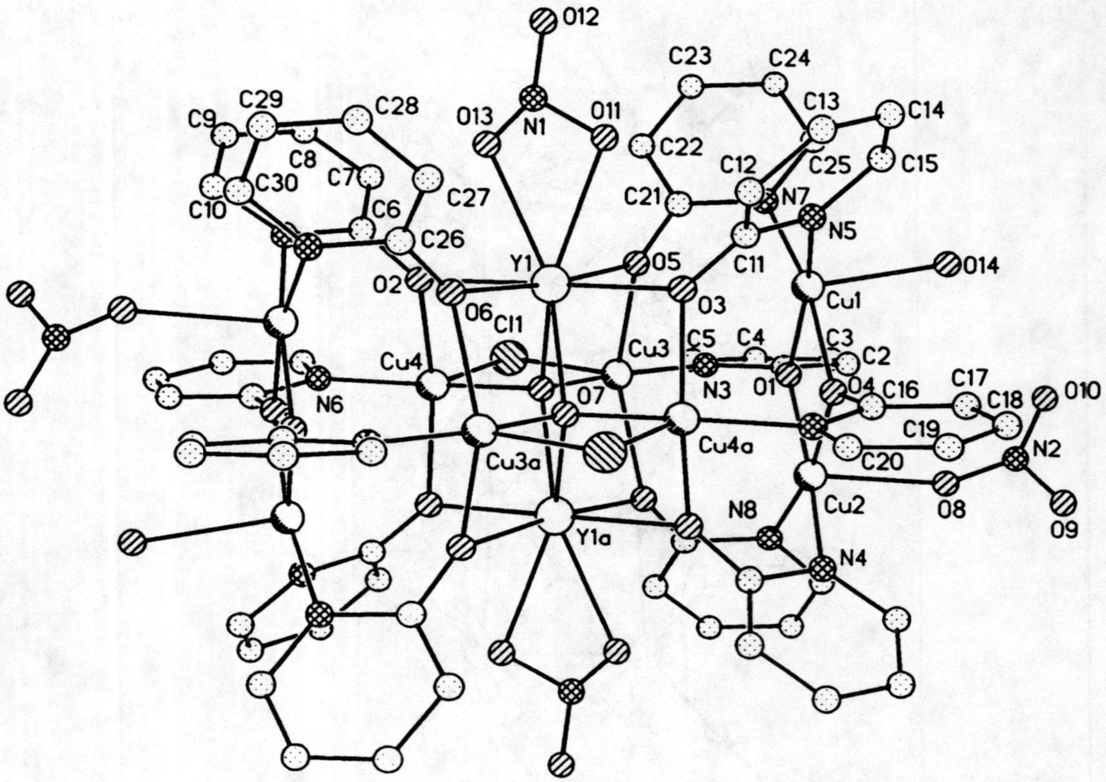

15. Structure of $Y_2Cu_8(pyO)_{12}O(Cl)_2(NO_3)_4(H_2O)_2$ [174].

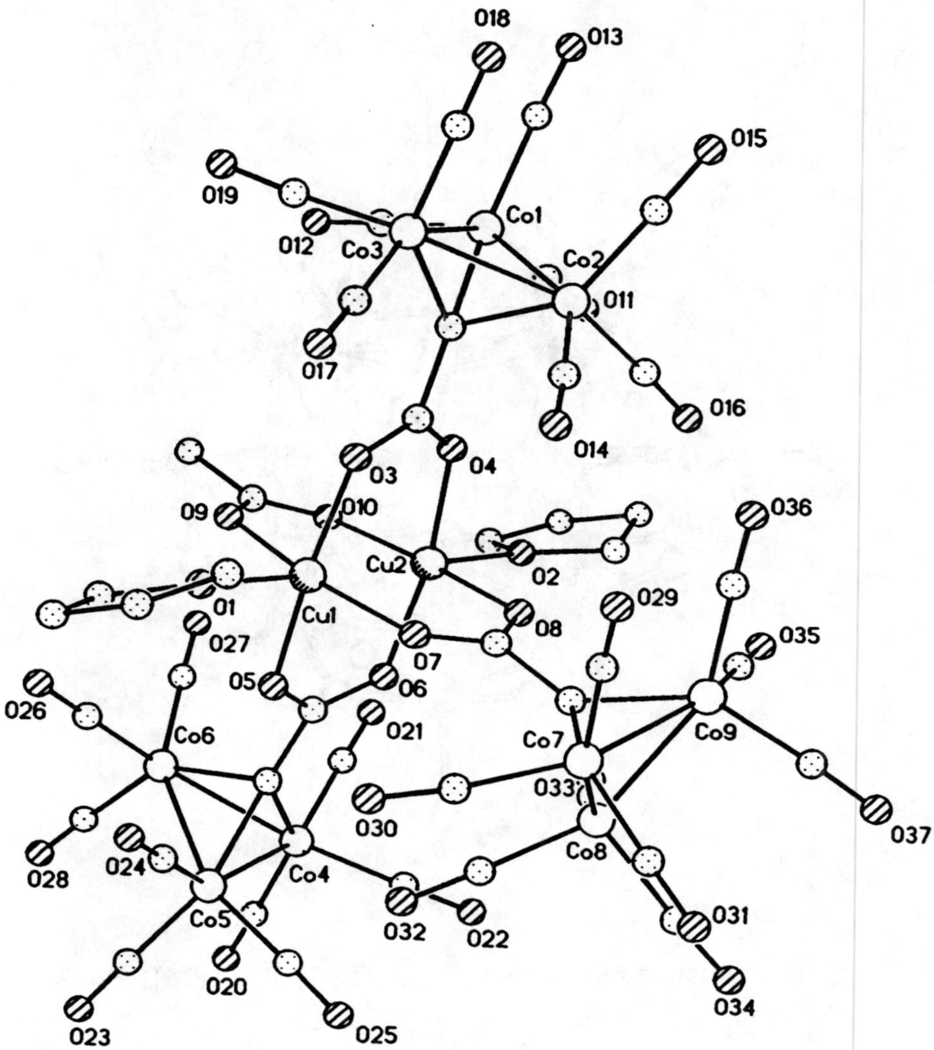

16. Structure of $Cu_2\{\mu\text{-}[(CO)_9Co_3(\mu\text{-}CCO_2)]\}_3(\mu\text{-}ac)_2(thf)_2$ [175].

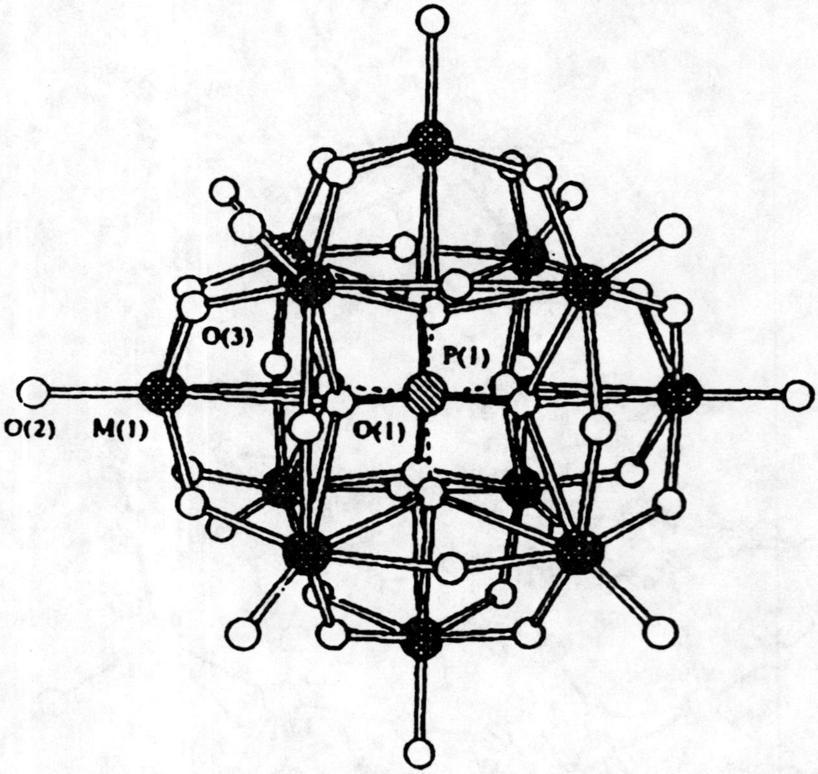

17. Structure of $[PW_3Mo_7Cu_2O_{38}(H_2O)_2]^{7-}$ [176].

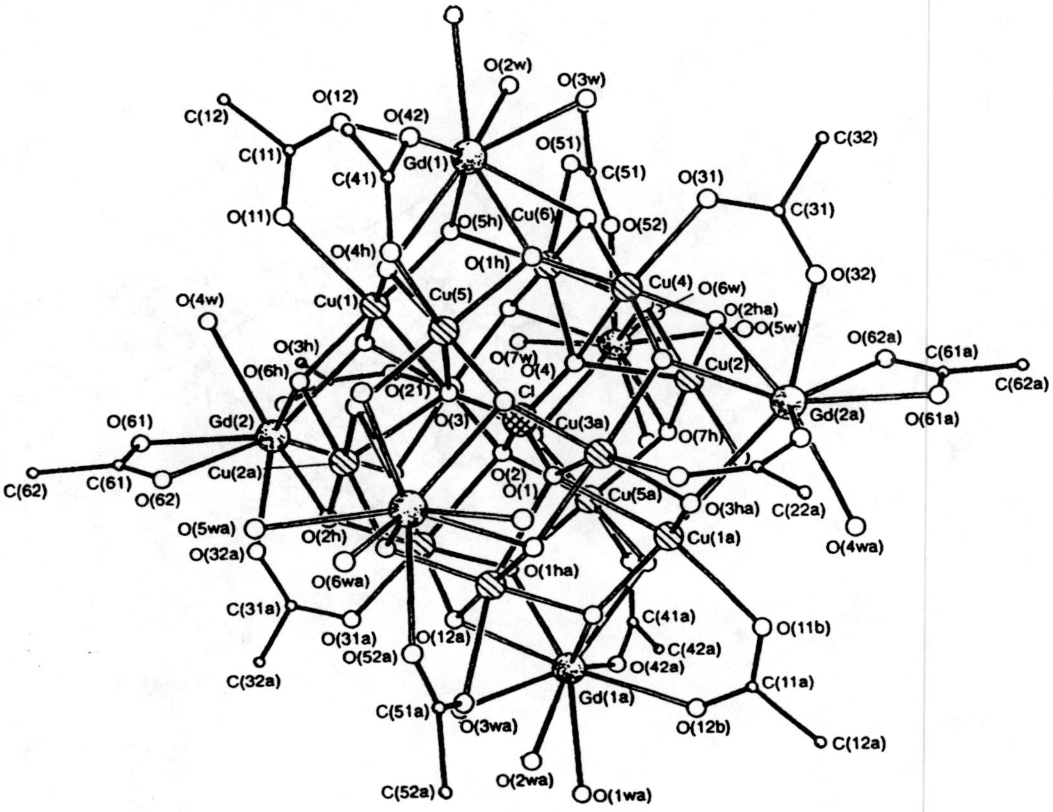

18. Structure of $Cu_{12}Gd_6(\mu_3\text{-}OH)_{24}(\mu\text{-}pyp)_{10}(H_2O)_{16}(\mu_{12}\text{-}ClO_4)]^{17+}$ [178].

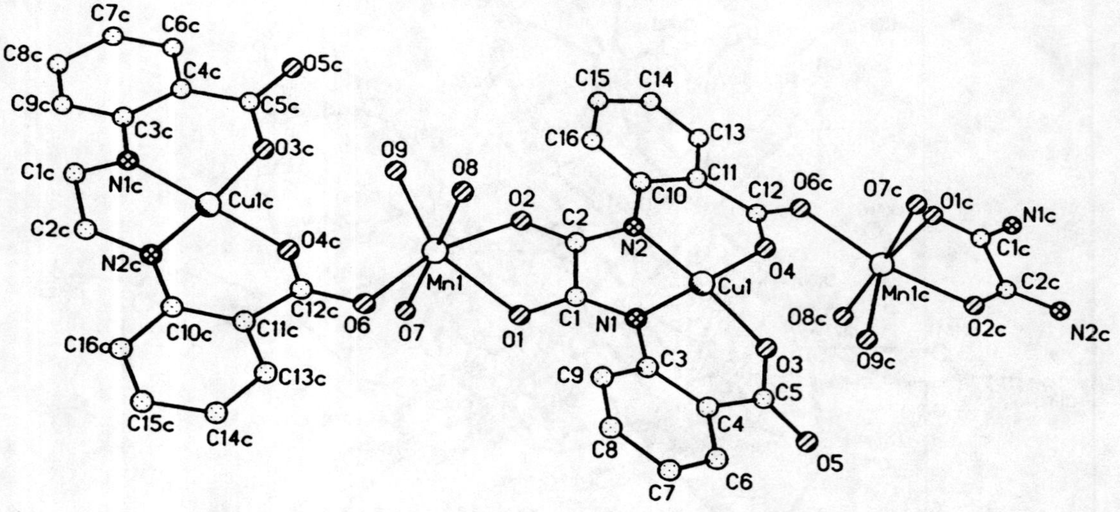

19. Structure of $[Cu(\mu\text{-obbz})Mn(H_2O)_3]_n$ [199].

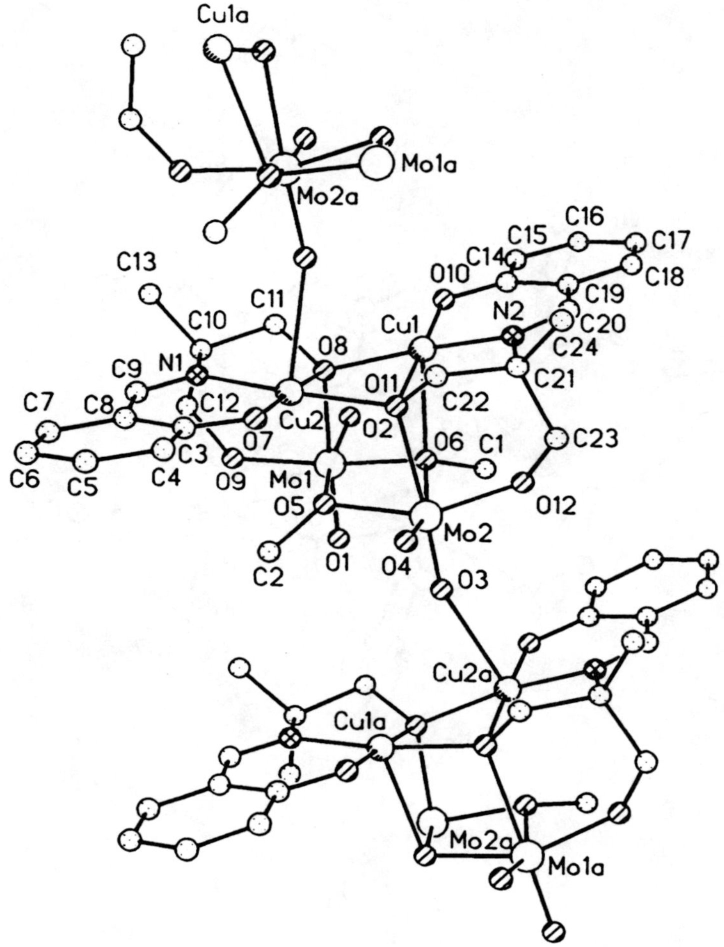

20. Structure of $Cu_2(MoO_2)_2(saldaph)_2(MeO)_2$ [212].

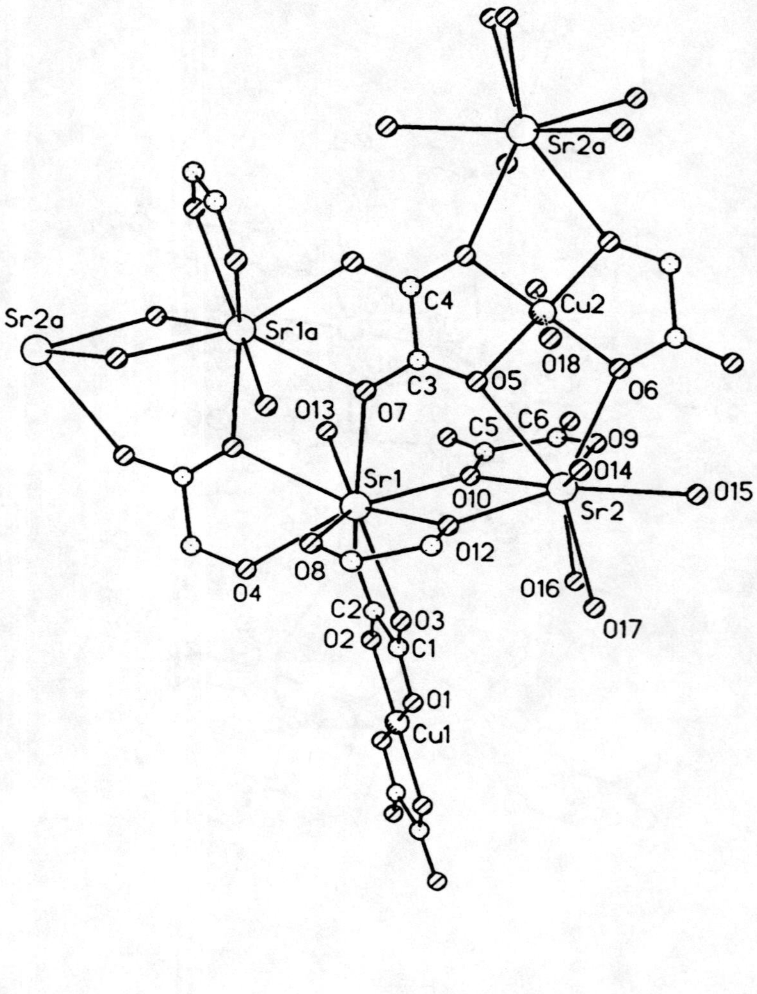

21. Structure of $\{Sr_2Cu(ox)_3(H_2O)_7\}_n$ [235].

INDEX

S

T

Y